INDIAN WAR CARTRIDGE POUCHES, BOXES AND CARBINE BOOTS

BY
R. STEPHEN DORSEY

Collectors' Library
P.O. Box 263
Eugene, Oregon 97440

Indian War Cartridge Pouches, Boxes and Carbine Boots
R. Stephen Dorsey

COLLECTORS' LIBRARY
P.O. Box 263
Eugene, Oregon 97440

Library of Congress Catalog Card Number 93-72940
ISBN 0-9631208-2-4
Printing Data: 10 9 8 7 6 5 4 3 2 1

DEDICATION

To the National Rifle Association
Which Protects Our Right
To Keep and Bear Arms

ACKNOWLEDGMENTS

This book is the result of the generosity of a number of collectors and students of the U.S. martial accoutrements. My thanks are offered to:

Dwight Clark
J. Edward Green III
Howard Hoffman
Zachary Leiner
Scott E. Meadows
Hayes Otoupalik
Michael Ward
Robert Yarmer
Rock Island Arsenal Museum
State of Oregon Library
University of Oregon Library

The author thanks Dr. Scott E. Meadows for allowing the use of photographs and data from his fine work, U.S. Military Holsters and Pistol Cartridge Boxes.

A special thanks is offered to my friend, Oscar Palmquist, who continues to spin marginal photos into gold and patiently provide computer order from manuscript chaos.

Any errors contained in this work are solely the responsibility of the author.

Table of Contents

The Hagner Cartridge Boxes

Hazen's Moveable Cartridge Loops

The McKeever Cartridge Boxes

Introduction

The Time

The year 1865 saw an exhausted South lay down its arms and the country as a whole looked west. The Army, however, was slashed–in size and budget–and found itself charged with protecting the nation as it moved tentatively into increasingly hostile territory beyond the Missouri River. The Army was expected to do all, be everywhere and to do it on Civil War leftover salt pork, beans and mountains of now-surplus equipage that ranged in quality from adequate to shoddy.

Technology had been one lesson the Army learned from the War. In 1865, the Spencer Carbine was in heavy use. Also in 1865, Springfield Armory brought forth the first regulation, metallic cartridge longarm, a conversion of the CW musket, also in .58 calibre. Admittedly too complex and underpowered, the 1865 Rifle Musket was quickly replaced in 1866 by an improved, converted musket. The Model 1866 was introduced with a new calibre, the .50/70 Government–a round to be used until 1873 in the Army and even later in state militias. The new converted muskets–as well as the older Spencer and the post-War converted Sharps Carbine–all fired cartridges with soft copper cases that required protection from denting and verdigris, though the soldier was now freed from worry about the fragility and sensitivity to moisture of the old paper cartridges. And, he no longer had to carry percussion caps.

The period of 1865-1870 was one of real frustration for field soldiers in the West. The aging Civil War musket and carbine boxes no longer served the needs of the far-flung Frontier Army. CW surplus accoutrements, uniforms, saddles and rations were either unfit as issued or quickly deteriorated in harsh Frontier service. Soft copper cartridges issued with CW musket boxes quickly battered themselves to unusability on marches, reducing the combat capability of troop units most in harm's way. And, slowly, the financially-strapped and ultra-conservative War Department and Ordnance Office in Washington began to respond to the reports of field commanders. The first type of response from Washington was to offer modifications to Civil War equipment to avoid the cost of development and production of all-new accoutrements. Early examples were the fleece linings sewn into CW rifle musket cartridge boxes and bored wooden block inserts for either the tin liners or the box cavity itself. In fairness, these early modifications did help somewhat but, increasingly, combat with a highly mobile and field wise Indian opponent called for different, more efficient and more robust equipment.

By 1870, the Army was in serious consideration of newer arms and accoutrements. Some patterns were suggested by soldiers in the field, others by inventors in private industry eager for a government royalty on their patents. For, as every inventor knew, the magnitude of a government adoption of his idea (if protected by a patent) was likely to pay far better than any such use on the commercial market.

By 1873, the Army had adopted yet another rifle and carbine, a new cartridge for them and had several, different metallic cartridge pistols in service. The Army was stretched even thinner against hostiles as far away as Fort Lapwai, Washington Territory, and the upper Yellowstone River. By 1873, soldiers were turning increasingly to the unauthorized use of hand-made and commercially-made "scouting" or "thimble" belts for field use, relegating cartridge boxes to garrison use. The tug-of-war between contending ideas and financial restrictions (as well as Ordnance Office recalcitrance) went on for years, joined by inventive, if often impractical, military and commercial accoutrement designers.

The use of the unwieldy, occasionally lethal (for its user) carbine thimble or socket and the shoulder sling preceded the Civil War. In 1859, the Army adopted the carbine thimble and seemed not to recognize the hazard and pain it and the slung carbine posed every time they were used. The hazard to the soldier thrown from his horse while the carbine was slung and in the socket finally was resolved–not by the Ordnance Office–but by an outcry from the field and direct intervention by General Sherman in the adoption of the Hartman Socket. It would be another five years before the weight of the slung carbine was finally relieved by the adoption of the 1884 Carbine Boot–some five years after a nearly identical boot, the Sumner Boot, had been submitted to the Ordnance Office from the 5th Cavalry. The 1884 Boot saw field modifications for several years, then was replaced entirely by the 1887 Carbine Boot, to better serve the Model 1884 Springfield Carbine with its more fragile rear sight.

Finally, like a great clock running down, it was all over. The 1890 massacre of Wounded Knee crushed the last vestige of native American hope for a return to the old ways and freedoms. The tribes that survived the white man's western movement–and many didn't–had been beaten, militarily quarantined and disenfranchised politically. Now dependent upon a largely disinterested and distant Washington for their survival, the western tribes no longer required the attention of large field units.

The cavalry and infantry continued to garrison many posts, camps and forts of the West, but their role became, increasingly, that of federal police and arbitrators of disputes between civilian elements of a land now connected by railroads and telegraph. As many old soldiers retired, they must have recognized that even the last days of the gunfighters had passed and the Old West was becoming a legend.

This Book

The period following the Civil War and extending through the advent of the truly "modern" arms of 1892 (the Krag Jorgensen Rifle and the Colt New Army Double Action Revolver) offers a tremendous variety of accoutrements for the collector and researcher. The array of cartridge pouches, boxes and carbine sockets and boots–depending upon the level of scrutiny–is both bewildering and, seemingly, endless.

This book is the result of the generosity of many collectors and researchers who offered their collections and expertise to establish the general outlines of the major pouches, boxes and carbine boots used or promoted for use in the 1865-1892 period. By and large, the book does not deal with Navy accoutrements, but does include several Navy and US Marine Corps pieces that are regularly seen in collections and at shows. And that leads to the basic purpose of this first edition: to show and discuss in historical perspective not only most of the pouches, boxes and boots likely to be seen by the collector, but many rare, additional items found in only a few collections throughout the US. A number of the items shown are possibly unique–the only surviving specimens. Some are rare and some, no less enjoyable and historic, are fairly accessible to the average collector. Truly, this is a hobby with a lot of enjoyment left in it.

The breadth of the artifacts shown here has allowed the author to present, for the first time, a true historical chronology of the development of the two major areas covered: cartridge pouches and boxes and carbine thimbles/sockets and boots. Many will be familiar, some will not. But all will be offered with a historical perspective long needed by the average collector.

As noted earlier, what is shown here are all of the historically important boxes, pouches and boots of the Indian Wars–and considerably more. What cannot be shown are the many prototypes, patent models, trial pieces, unrecorded arsenal modifications and a large number of state militia and military school variations of common and uncommon accoutrements. Some of these are recorded, but are unknown as artifacts, some are merely repetitious and some esoteric variations were not made available by their owners. Significantly, though, this has had no impact on the step-by-step chronology of development and use of the key pouches, boxes and boots, presented here. As new artifacts are identified and made known, they can be woven into the story in their rightful place.

Organization

The first six sections deal with the major cartridge pouches and boxes of the period:

Civil War Reissued and Converted Cartridge Pouches and Boxes

The Dyer Cartridge Pouches

The Hagner Cartridge Boxes: Regulation, Militia and Unknown

The Hazen's Moveable Cartridge Loops

The McKeever Cartridge Boxes: Regulation, Experimental and Militia

Patented, Commercial and Experimental Cartridge Boxes

Carbine Thimbles, Sockets and Boots is the last major section of the book, and it covers the development of these items from the time of the Dragoons through the 1887 Carbine Boot.

Where necessary to make distinctions between similar items or to describe unusual and less well known items, the following headings have been provided:

Calibre
Capacity
Measurements
Colour/Materials/Construction
Markings
Comments

Under Measurements, the abbreviations shown were used:

OAL–Overall Length
OAW–Overall Width
OAT–Overall Thickness
OAH–Overall Height (where more descriptive)

Measurements did not include finials nor belt loops.

Civil War Reissues and Modifications

Civil War (Percussion) Cap Pouch

Calibre: NA

Capacity: Could carry at least 200 rifle musket caps and many more pistol caps. Also carried a cone pick, captive in a leather loop in a back corner of the box.

Measurements: There are minor variations in these cap pouches but the specimen shown measures:

OAH–	3 5/16"
OAW–	3 9/16"
OAT–	1 5/8"

Colour/Materials/Construction: Uniformly black, these stiff leather pouches had an outer cover/flap with integral closing tab and an inner flap. (Photo 2) Beneath the inner flap, the opening to the pouch cavity was filled with wool from a strip of sheepskin sewn on the back of the box. (Photo 3). The interference of the wool in the mouth of the pouch cavity prevented caps from bouncing or jostling out. As shown in Photo 3, the inner flap had two attached ears, one each, sewn to the outside ends of the inner flap. These ears further protected the percussion caps from rain or debris entering the pouch from the sides.

The back of the box had two

Photo 1.

Photo 2.

belt loops for the standard leather waist or sabre belt attached with stitching or rivets and stitching to the back of the box.

Markings: These pouches, made over a long period of time and by a myriad of producers, have a wide variety of markings. Almost all are stamped with an inspector's cartouche (usually a large oval containing inspection data) and many are stamped on the inner flap with a maker's name and city. A commonly-seen, flower-like device is not an informational mark but merely a mark left by a tack used to secure the work to a board for stitching by the leatherworker.

General Order No. 60, dated June 1872, stated that, "The Infantry cap-pouches, for the present, will be used." (For pistol cartridges.) This use, due to the tremendous surplus of the pouches left from the Civil War, continued until the early 1890s, from photo evidence of that time frame. Note that the order made no mention of modification of the pouches (wool or inner flap removal, etc.) so these refinements were left to the decisions of the field and account for the great variety of modifications seen.

The most common modifications of the cap pouch to cartridge pouch were removal of the sheepskin strip at the top of the box and removal of the inner, leather flap. A pouch incorporating both of these changes (field modifications) is shown in Photo 4.

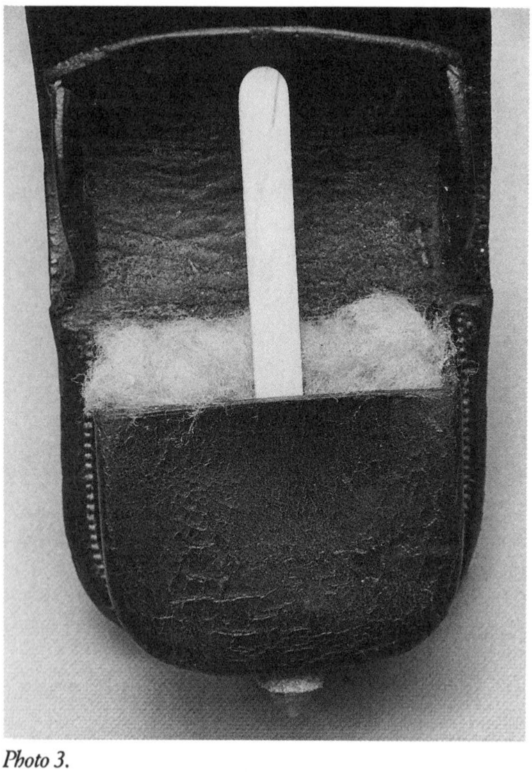

Photo 3.

Photo 4.

Field Modification of CW Cap Pouch With Loop for M1876 Cartridge Belt

The loops of the Civil War cap pouch had been designed for use on the leather sabre or waist belt of the period. When the Army finally adopted a cloth cartridge or "prairie" belt for its Frontier forces, the belts were too wide to accommodate the cap pouch belt loops. A soldier wishing to carry his pistol cartridge pouch (cap pouch) on his cartridge belt was required to modify the pouch loops to fit the new belt–or, continue to wear the pouch on his leather sabre belt.

Photo 2 shows the back of two cap/pistol cartridge pouches. The pouch on the left is in the regulation CW configuration that was, in 1872, adopted for use as a pistol cartridge pouch. The pouch on the right is a high quality field modification of a similar cap pouch, in which a much longer and wider single belt loop was substituted. (The stitching marks from the earlier loops clearly show.) This longer, wider loop allowed the pouch to be carried on the M1876 Prairie Belt and it is shown mounted on such a belt in Photo 1.

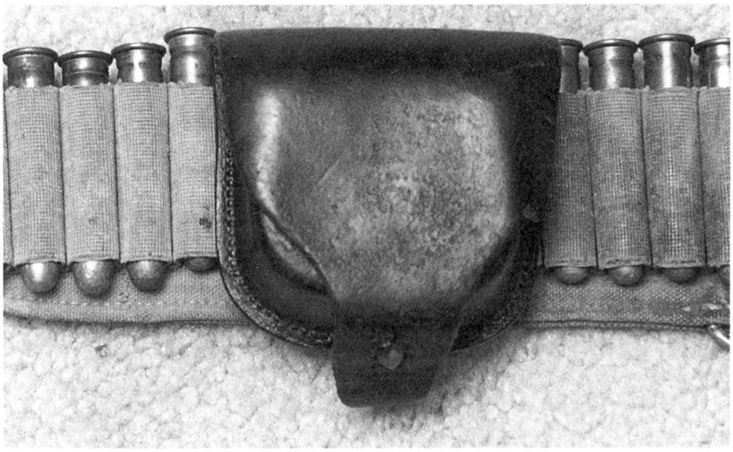

Photo 1.

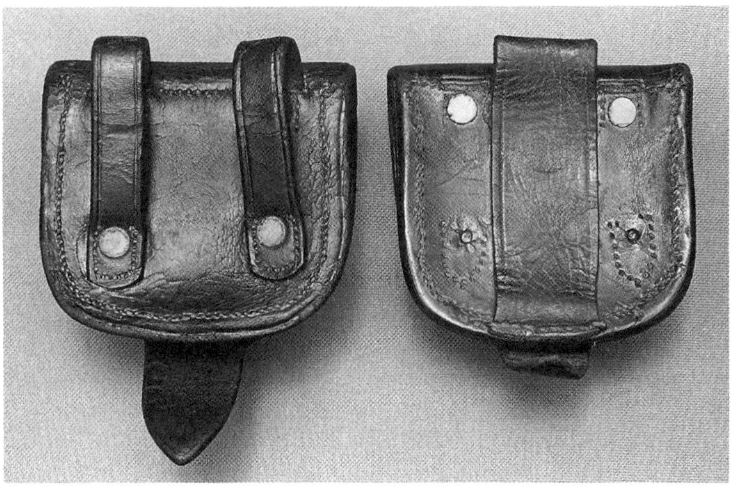

Photo 2.

Modification of Cap Pouch With Wooden Block

As shown in Photo 1, this pouch was the standard CW Cap Pouch before modification. The conversion consisted of removal (by cutting) of the inner flap and insertion into the cavity of a bored wooden block.

Superficially, the modified pouch would appear to accept any .45 calibre pistol round, but in reality, only the 1875-era .45 Schofield round will allow the flap to be pulled over the finial. Further, this block has also been seen in the M1875 Pistol Cartridge Pouch. These facts, and the high, uniform quality of these blocks, leads the author to think that either the blocks were arsenal-made and made available to troop units for their own conversion, or were both made at an arsenal and actually part of a conversion process there.

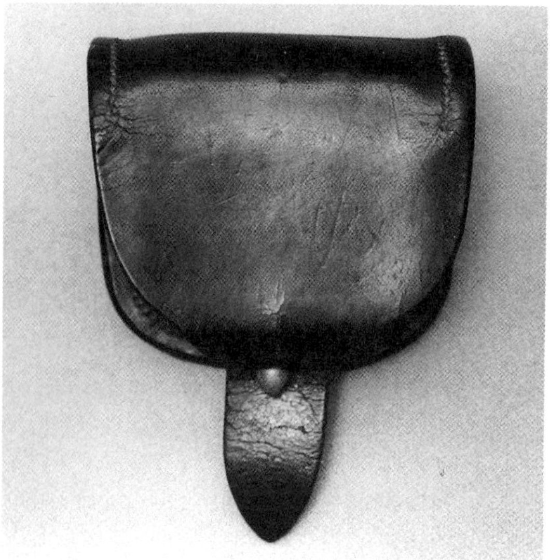

Photo 1.

Photo 2.

1875 Modified Pistol Cartridge Pouch

Calibre:.50 Remington Pistol, .44 Smith and Wesson, .44 Colt, .45 (Long) Colt and .45 Smith and Wesson Schofield. The pistol which accompanied the 1872 Cavalry Brace System was the .50 calibre single shot Remington Army Pistol ("Rolling Block"). This obsolescent weapon was quickly dismissed from the service, being replaced by the S&W No. 3 "American" Revolver, the M1873 Colt and the S&W Schofield. In addition, the .44 Colt calibre Richards Conversion of the M1860 Colt was in service for much the same period. All of these cartridges could be carried in the 1872 Pistol Cartridge Pouch.

Capacity: The size of the cartridges (discussed above) varied greatly and the number carried in this pouch depended upon the calibre chosen. As many as 16-18 .50 Remington rounds could be carried and as few as approximately 12 .45 (Long) Colt rounds.

Measurements: Identical to the standard Civil War Cap Pouch. Approximately:

OAH –	3 3/8"
OAW–	3 1/2"
OAT–	1 7/8"

Colour/Materials/Construction: The 1875 Pouch was originally a standard, new-condition Civil War Cap Pouch of heavy black leather. It had two standard belt loops on the back. This standard box was then modified at Watervliet Arsenal in the following manner:

1. Wool and cone picks removed from cavity.
2. Inner flap removed by cutting at juncture with box back.
3. Two leather "ears" (Photo 2) sewn over the outside edges to reduce the likelihood of cartridges jolting out when the pouch was open or closed.
4. Flap or cover deeply wet-embossed with oval containing the letters "US". This embossing is typically done over the Civil War inspector's stamp (Photo 1).
5. Inspector's stamp for Watervliet Arsenal modification stamped diagonally on flap tip: **"A.R. SMITH."**

Markings: None, other than described above.

Comments: A soft leather, gusseted pistol cartridge pouch had been adopted by the 1874 Ordnance Board. This pouch, a reduced-size version of the 1874

Photo 1.

Photo 2.

Photo 3.

Dyer Carbine Cartridge Pouch, was recommended, but was not approved by the Ordnance Department, due to the great surplus of CW cap pouches in inventory. However, in 1875, Col. P.V. Hagner, Commanding Officer of Watervliet Arsenal, wrote the Chief of Ordnance regarding an order for 1,000 pistol cartridge pouches. In this correspondence, Hagner noted that the order could be filled with CW cap pouches, **"altered to close agreement with the prescribed [Note: 1874 type] pistol pouch..."**. This pattern was approved in December of 1875 and is believed to be the pouch shown. (Photo 3)

The 1875 Pouch is seen both without the wooden block (most common) and with it. Those with the wooden block were assembled in such a way that the "ears" were stitched over the ends of the block so that it cannot be removed. The block is bored for six rounds of .45 Schofield c.1875, and cannot be closed on six rounds of the longer .45 Colt round. This leads the author to believe this is a late production of the pouch.

The collector should be aware that the standard M1875 box has, unfortunately, been forged by late 20th century fakers. These late specimens have the embossed "US", but are not inspected, nor is the quality of the work on the "ears" very good. Comparison with an original will clearly show up the discrepancies.

Field Field FieldModification of Cap Pouch With Carrying Strap and Buckle

Sometime in the late Indian War period, it appears that the 10th Cavalry modified a number of pistol cartridge pouches (re-issued CW cap pouches) for carriage by a short strap and buckle arrangement. The workmanship of this modification is serviceable, but is not of depot or arsenal quality. Of the three specimens examined, there were minor variances in workmanship –indicating, probably, manufacture by a company saddler.

These modified pouches apparently were hung by a flap or tab descending from a sabre or cartridge belt or from the face of a modified revolver holster. Such a modified holster has not been seen by the author, but the arrangement could have been as simple as a descending tab riveted to the front body of the holster or belt body. Holes in the tab would have engaged the tongue on the buckle of the pistol cartridge pouch.

At least one variant to this mounting is known. It consists simply of a cap pouch/pistol cartridge pouch riveted to the lower front of the revolver holster. Two examples of this latter type of modification are shown on .38 Colt Revolver holsters carried by 10th Cavalry troopers in 1894. (Montana Historical Society photo shown on page 95 of Vol. II, Firearms of the American West, by Garavaglia and Worman).

Though the addition of a loaded pistol pouch would introduce a pendulum-like swinging of holster and pouch, the easy access of the revolver ammunition may have overshadowed any such drawbacks.

This type of cartridge pouch/holster modifications obviously continued in military service into the mid-1890's. It was only in 1896 that the Army adopted a new pistol cartridge pouch that would serve either .45 or .38 calibre revolvers.

Photo 1.

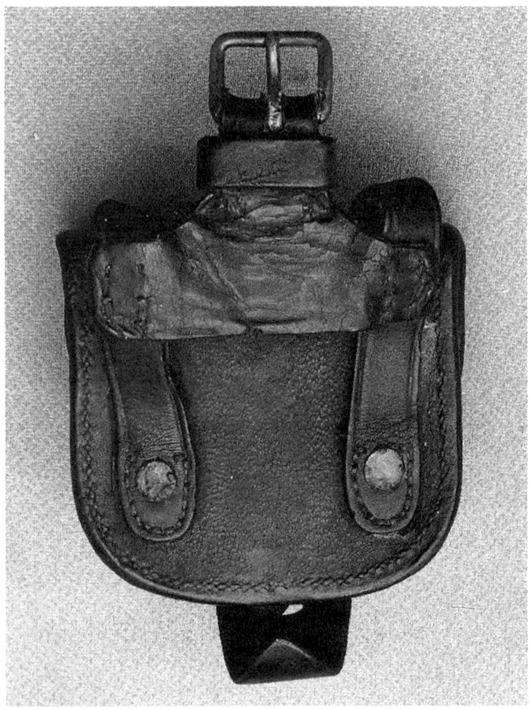

Photo 2.

Field Modification of Cap Pouch with Long Strap

In 1878, Rock Island Arsenal manufactured 100 experimental Second Pattern M1876 "Prairie" or Cartridge Belts that differed primarily from the standard pattern in having an inside leather strap. These were stamped on the reverse **"R.I.A. EXPERIMENTAL."** San Antonio Arsenal received permission to modify a number of standard M1876 Cartridge Belts in a similar fashion.

These special cartridge belts lent themselves to carrying not only the bayonets for which they had been designed, but other accoutrements that could be modified to utilize the inner strap. The accompanying photo shows a CW cap pouch modified with a long loop to fit on the inner leather strap of either the R.I.A. experimental belt, or the slightly wider San Antonio-modified belts.

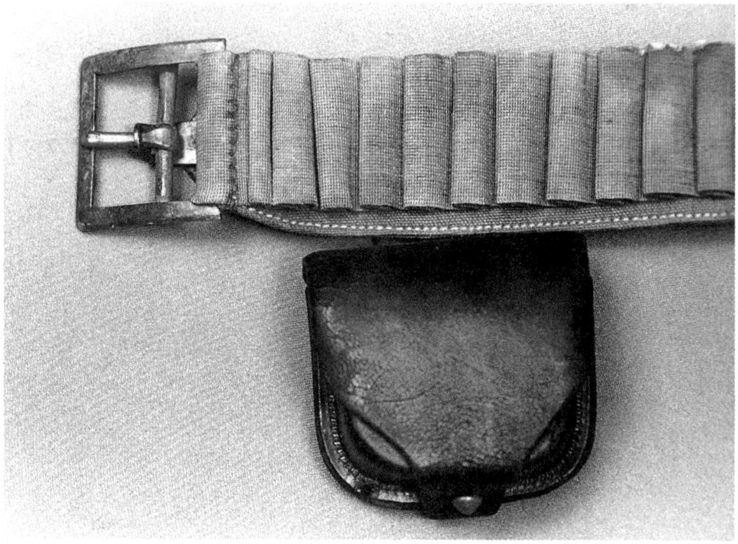

Photo 1.

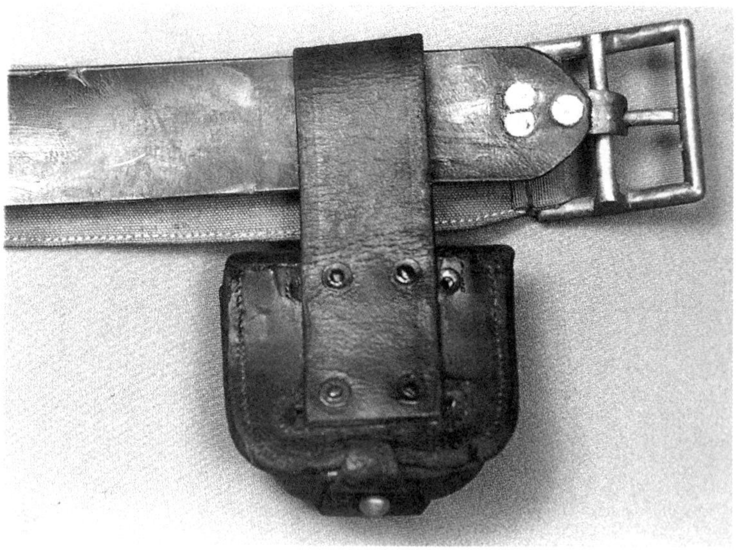

Photo 2.

Civil War Pistol Cartridge Box

Calibre: A generic box to protect and carry six-round packets of .36 or .44 calibre revolver cartridges.

Capacity: Varied with size of box–from two to six 6-round packets of cartridges.

Measurements: These boxes varied in size a great deal, though they are of almost identical construction. The pictured box has the following measurements:

OAL–	Flap	5 3/4"
	Box Body	4 5/8"
OAH–	Flap	3 3/4"
	Box Body	3 1/4"

However, the largest size in the author's research collection measures as follows:

OAL–	Flap	7"
	Box Body	4"

Depending upon the calibre of the cartridge and the method of packaging, the wide range of intermediate sizes between the two specimens listed above could carry from two to six packets of ammunition. It is the author's opinion that the many, slightly different sizes of this box more reflect commercial manufacturers' variations, rather than specified sizes for specified calibres and cartridge packets.

Colour/Materials/Construction: These boxes were manufactured in stiff collar leather. They were all originally black but some, due to dye deterioration, have turned brown over the years. Each box has two belt loops riveted and sewn on the back. The closing tab on the front was first (pre-War) attached with a straight line of stitching, but later was more securely attached with a rivet and a curved line of stitching. The brass closing finial was centered on the box bottom.

Each box had a central cavity and an implement pocket at the front of the box, created simply by adding a full leather partition at the front of the box. This pocket would carry the L-tool, a combination cone wrench and screwdriver. There was no inner flap as found on cap pouches and rifle and rifle musket cartridge boxes.

Markings: These boxes vary from unmarked–as the specimen pictured in Photos 1-3 –to well-marked by both manufacturer and inspector. Some are marked with a script "US", centered on the cover, but are still felt to be of standard Civil War production. None were ever intended to carry a brass plate, but a few are seen so modified.

Comments: When fully loaded with packets of cartridges, these stiff, deep boxes were very difficult to use. The cartridge packets were wedged deep in the box body and the stiff sides resisted the wearer's fingers as he attempted to grasp one or two packets.

Photo 1.

Photo 2.

It is the author's opinion that, though a regulation pistol cartridge box as long as percussion revolvers were in service, these boxes were unpopular and largely unused. This is borne out by the excellent condition of the many specimens seen. At best, they were probably never fully tightly loaded, so that the packets carried could be moved and grasped quickly.

These boxes probably saw some use in the Regular Army and state militias into the 1870s.

Photo 3.

Modification of Pistol Cartridge Box With Wooden Block

A very unusual and rare modification of a Civil War Box for early Indian War use.

The Spencer Carbine in 56/.50 calibre served for some time on the Frontier. The Spencer was rapid fire, robust and relatively hard-hitting. First used by the Frontier army cavalry, the Spencer was also provided to state militia cavalry such as Colorado, Arizona and Oregon.

Despite the availability of Blakeslee Quick Loaders, they appear not to have been issued on the Frontier, leaving soldiers to utilization of on-hand boxes pouches or manufacture of belts to carry Spencer rounds. This CW box contains a machine-made wooden block, bored with seven holes for the 56/.50 round. This block was not likely manufactured in the field, but probably was arsenal or depot made. Though offering only a seven-round capacity, it was compact and could have been carried on the waistbelt or carbine sling, singly or in multiples.

The "U.S." on the cover has no Indian War significance. It is a variant CW manufacturer's embellishment.

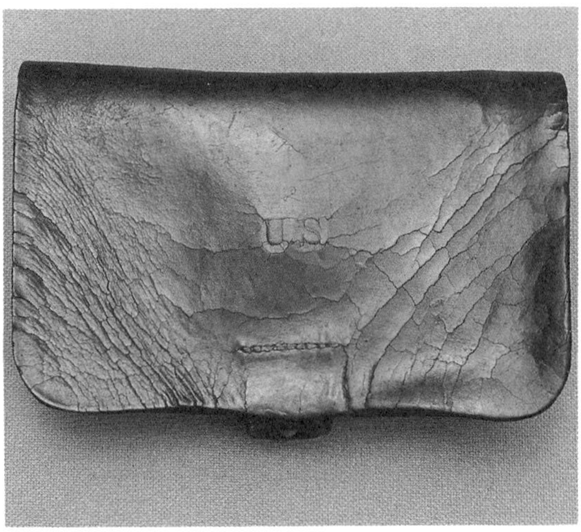

Photo 1.

Photo 2.

Civil War Carbine Cartridge Box

Calibre: These boxes were issued with bored wooden block inserts that allowed issuance with any issue carbine–just by changing the block. Normal calibre range: .50 to .56

Capacity: 18-20 rounds, arranged in two rows

Measurements:

OAL–	8 1/8"
OAH–	4"
OAT–	2 1/4"

Colour/Materials/Construction: The black, bridle leather box shown is a late-CW pattern, manufactured without two shoulder strap retaining buckles on the bottom of the box. As shown in Photo 3, there is an elongated implement pocket on the inner face of the box body as well as an inner, protective flap. The back of the box has two tall belt loops that will allow it to be worn on the waist belt or carbine sling.

Markings: These boxes are normally marked with a maker's name and an inspector's mark. The latter on the cover, usually.

Comments: As mentioned before, the carbine cartridge box had great versatility–just by changing the wooden block (Photo 4)–so it continued to be widely used after the War and on the Frontier. In the immediate post-War period, the linen cartridge .52 cal. Sharp's Carbine, the metallic cartridge .50 cal. Spencer Carbine and the Starr Carbine were all served by the same box. Later, the cartridge box served the metallic cartridge conversion Sharps as well as the Spencer Carbine. By 1870, the Dyer and other pattern boxes had supplanted the older carbine cartridge box, and what use it continued to see was in state militias.

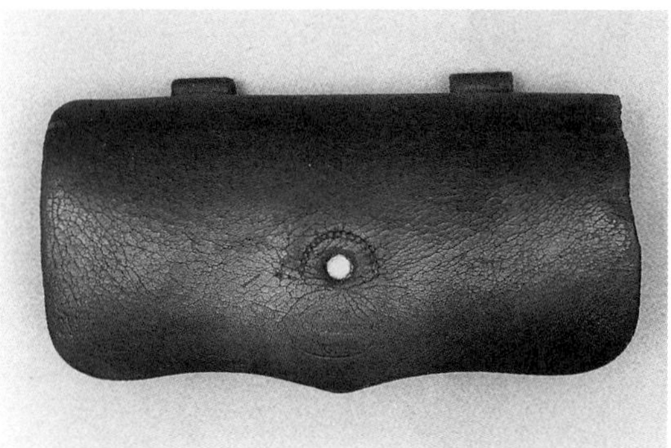

Photo 1.

Photo 2.

Photo 3.

Photo 4.

Fleece-Lined Model 1855 Rifle Musket Cartridge Box

Calibre: Initially, possibly .58 Centrefire for the M1865 Springfield (First Allin Conversion) and the 56/.50 Calibre Spencer Carbine.

Capacity: Dependent upon calibre, but at least 25 rounds or more.

Measurements:

OAL–	8"
OAW–	7 1/4"

This cartridge box is the standard M1855 Rifle Musket Cartridge Box with two modifications: The twin tin liners/boxes have been removed and the interior of the cartridge box has been fully lined with white fleece. As shown in Photo 2, the fleece liner was stitched in place in a semicircle on the underside of the inner flap and across the front, upper lip of the box. The narrow side panels were also stitched in place at the top.

Markings: The lower, center edge of the box flap is stamped with the inspector's initials **"F.M.M."** These were applied at the arsenal upon the inspection of the modification.

Comments: A very early and desirable IW cartridge box, this is probably the first rifle musket cartridge box modified after the Civil War. It is known, however, that immediately after the Civil War, Frontier troops were simply issued the regulation rifle musket cartridge boxes with tin liners–much to the consternation of soldiers and officers such as MAJ. Anson Mills. The battering of the soft copper rounds and the loud rattling attendant to the tin-lined boxes led to the modification of Civil War boxes such as this one.

These and the unmodified boxes are known to have been carried on the original shoulder sling sometimes, but they certainly could have been worn on a belt, if desired. This fleece lining seems to have been largely confined to M1855 Cartridge Boxes, not the M1864 (embossed U.S.) boxes.

This early, modified cartridge box is quite desirable and scarce. They are proper in display with several early, post-War longarms.

Photo 1.

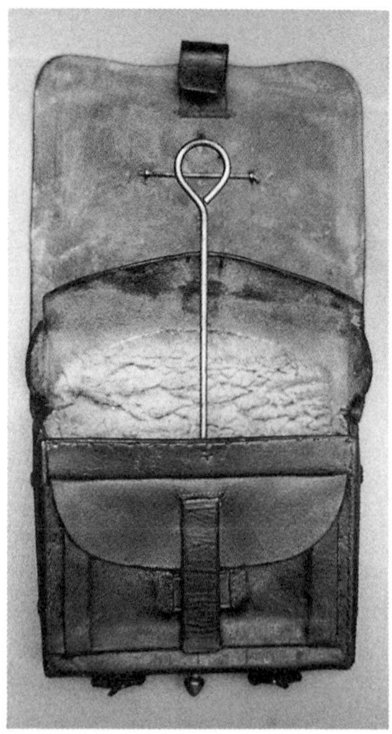

Photo 2.

Wooden Block Modification of Model 1864 Rifle Musket Cartridge Box

Calibre: .45/70 Government

Capacity: 20 rounds in two rows of 10 rounds each

Measurements:

OAL–	8 1/4"
OAW–	7"

Colour/Materials/Construction: This is a standard M1864 Rifle Musket Cartridge Box with its two tins removed and replaced with a bored wooden block. The block is not tacked in, and sits high in the box cavity.

Markings: The original Civil War manufacturer's name of "S.M. Young Newark N.J." is stamped on the back of the box. An original CW inspector's stamp shows on the face of the box.

Comments: A desirable, early conversion cartridge box that may have been issued with a shoulder sling or may have merely been worn on the belt. This box is an example of the cartridge boxes actually in use while the Army Ordnance was attempting to adopt a superior cartridge box through a series of boards and trials.

The financially strapped post-War Army found such conversions easy and low-cost in comparison to cartridge boxes such as the Dyers and Hagners, which had to be manufactured from scratch.

Photo 1.

Photo 2.

.58 Calibre Rimfire Musket Cartridge Conversion Block For Rifle Musket Box Tin

Comments: Shown here is a specimen of a scarce, early conversion of the standard Civil War rifle musket cartridge box to carriage of the first post-War cartridge, the .58 Rim Fire Musket round.

The wooden block is obviously of arsenal manufacture and made for the Regular Army inasmuch as the M1865 Springfield Rifle Musket (1st Allin Conversion) did not see militia issue. The block is a simple drop-in unit that could have been issued simultaneously with the muskets to troops already equipped with the Civil War rifle musket cartridge boxes.

Photo 1.

The Dyer Cartridge Pouches

Overview

In 1870, the Army was coming to grips with the magnitude of the task of taming the post-War Frontier. Clearly, the war on the plains was different: distances were greater, more arid and much, much further from basic supplies than in any other war in our history. Civil War equipment was proving inappropriate and inadequate–as well as wearing out faster than anticipated. The combat troops engaged on the Frontier were overwhelmingly cavalry–but thinly spread. Cartridge guns had replaced percussion arms, and this and other factors was making a serious review of cavalry equipment necessary. In this historic perspective, Chief of Ordnance Alexander B. Dyer offered the first of his developmental cartridge pouches. Alexander Byrdie Dyer was born in North Carolina and graduated from West Point in 1837. Commissioned in the artillery, he transferred to ordnance in 1838. He served in the Mexican War, being brevetted captain for his part in the battle of Santa Cruz de Rosales. He became brigadier general and Chief of Ordnance in 1864 and was brevetted major general in 1865. He served as Chief of Ordnance until his death in 1874.

The first Dyer Pouch was the now-rare Sling Pouch "E", one of seven concept cartridge pouches submitted to an 1870 Army review board. One hundred were manufactured and issued around the country for test. A number saw combat usage with Spencer and Sharps (cartridge) Carbines against the Modocs in the famous Lava Beds War.

The Sling Pouch "E", as will be seen in the following coverage, was a large, soft-leather, flapped pouch with both belt loops and a combination carbine sling and shoulder strap, somewhat narrower than the Civil War model. This sling was equipped with a standard pattern (carbine) sling swivel.

In 1872, the Army adopted for test a cavalry brace system that incorporated a belt-mounted Dyer Carbine Cartridge Pouch, reminiscent of the earlier pouch, but slightly smaller and fleece-lined. This carbine cartridge pouch was liked by a majority of the cavalry officers polled, but the '72 Brace System was roundly disliked and the brace system was dropped as a cavalry concept.

In 1874, Dyer had two new pouches for Army consideration. The first was a new carbine cartridge pouch–downsized, fleece-lined and with a brass escutcheon over the closing hole. A second, extremely rare Dyer Pouch was pictured (with the larger 1874 Carbine Cartridge Pouch) in Ordnance Memo No. 18 for the cavalry equipments of 1874. This Dyer-pattern pistol cartridge pouch was never made in quantity. Before quantity production was begun, cost-conscious ordnance officialdom overturned the approving Board's recommendations on the pistol pouch and continued the use of the Civil War Cap Pouch for pistol cartridges (begun earlier under the 1872 General Order No. 60). It is noted that minor variations of Dyer-type pistol cartridge pouches are reported, leading the author to think that, perhaps, the period between 1872 and 1874 saw the production of several, trial transition pouches. These variations include minor loop variations, variations in the shape of the pouch and variations in the shape of the flap.

The Model 1874 Dyer Carbine Cartridge Pouches saw wide distribution but were largely eclipsed for combat use by the "scouting" or "prairie belt." The "prairie belt", first those unofficially manufactured and worn, then a series of official models, saw uniform cavalry combat use for decades. The M1874 Dyer Pouch, however, seems to have been dropped from Army service by the mid-1880s.

1870 Dyer Sling Pouch "E"-Carbine Cartridge Pouch

Calibre: Designed for .50/70 Government and 56/.50 Spencer rounds

Capacity: Sixty .50/70 Government; Estimated seventy 56/.50 Spencer rounds in large pocket.

Measurements: Pouch

OAW–	7 3/4"
OAH–	6 1/2"
OAT–	2 1/2"
Sling OAW–	1 5/8"
Buckle–	1 9/16" x 2"

Colour/Materials/Construction: This soft, well-made black leather pouch is designed to either be carried by its shoulder sling or by its two belt loops (sling detached.)

The pouch body has wide, soft gussets that allow the pouch to expand and contract as the cartridge load required. The front/face flap fits over a brass finial and has two integral side "ears" that are rain and debris protection on the upper side.

An inner revolver cartridge pouch is delineated on the face of the pouch body by a large semi-circle of stitching. That smaller pouch also has a soft leather closing flap, which uses the common closing finial for the main flap.

The main cartridge pouch is lined with white sheep's wool to protect the soft copper cartridge cases.

The sling is adjustable by engagement of a single brass hook into holes spaced along part of the sling body. The brass buckle is captive on one end of the sling. All pouch fittings are brass. Full length, the sling measures 78 1/2" long.

Though not shown in the photos, the sling itself carried a Civil War-type carbine sling swivel of a type narrower than the CW swivel, but otherwise identical. In this way, the Sling Pouch "E" functioned both as a carbine ammunition pouch as well as a carbine sling.

Markings: As shown in Photos 1 and 2, the pouch is stamped **"U.S. WATERVLIET ARSENAL"** in three lines, centered on the main flap.

Comments: This sling pouch was designed by the Chief of Ordnance Alexander B. Dyer and submitted to Army equipment trials in 1870. One hundred specimens are documented as having been manufactured and distributed to several arsenals and a depot for troop issue and test: 25 to Leavenworth Arsenal, 24 to Omaha Depot, 24 to San Antonio Arsenal, 25 to Benecia Arsenal and two remained at Watervliet Arsenal.

The sling pouch, doubtless, saw troop test in each of these widely-separated regions, but the best-known (and photographed)

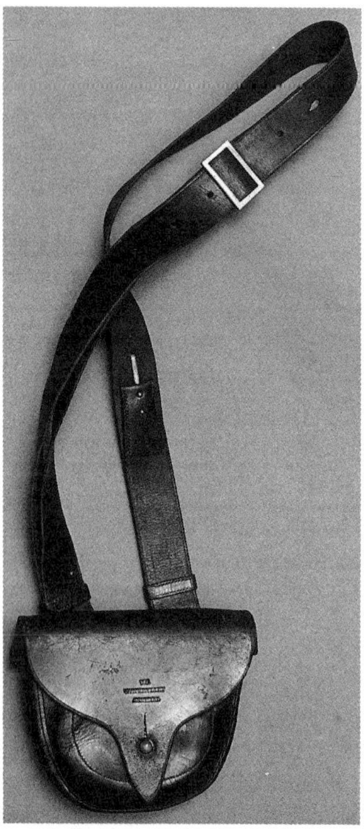

Photo 1.

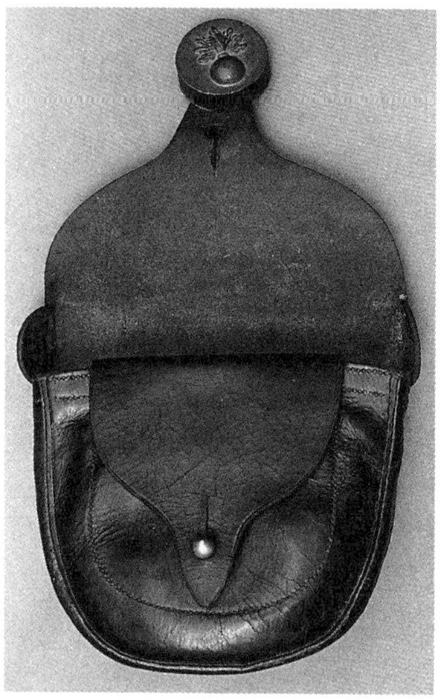

Photo 2.

use was in the hands of the First Cavalry in the famous Modoc War in the Lava Beds Campaign.

Appearing quite simple, the Dyer Pouch was very utilitarian and sturdy. It admitted use with a wide variety of cartridges, could be worn on shoulder or belt and served as ammunition and maintenance tool holder as well as a carrier for the army issue carbine (on the snap swivel.) It was, apparently, popular and led to the design refinements found in its successor, the M1872 Dyer Carbine Cartridge Pouch.

The Dyer Sling Pouch "E" is a rare and historically-important IW accoutrement.

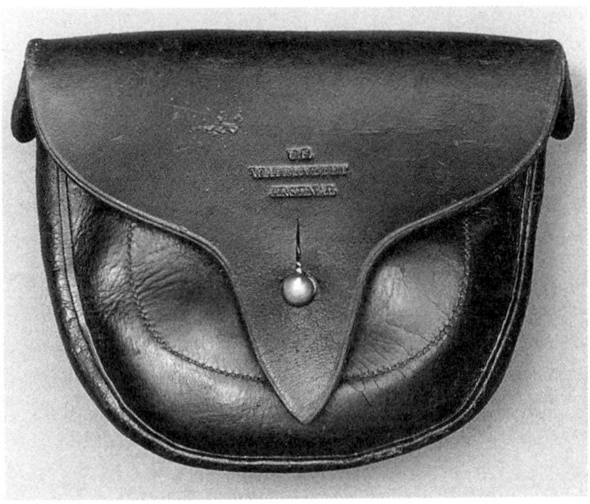

Photo 3.

Photo 4.

1872 Dyer Pistol Cartridge Pouch

Photo 1.

Photo 2.

Calibre: .50 Remington, .44 Colt, .44 Smith and Wesson

Capacity: Unknown, but at least 12 rounds

Measurements:

OAH–	3.84"
OAW–	4.1"
OAT–	1.075"

Colour/Materials/Construction: Following closely the contemporary construction of the carbine cartridge pouches being considered, this pouch is also a soft, gusseted collar leather pouch with belt loops extending under the pouch (to accommodate the fittings on a cavalry sabre belt).

The pouch was originally fleece-lined, and has a brass finial and rivets. Two rivets secure the "ears" on each side of the opening. These ears were to inhibit cartridges from jostling out when the pouch was opened.

The closing tab is, typical to the period, without a brass escutcheon.

Markings: None, but probably manufactured at Watervliet Arsenal.

Comments: This truly rare pistol cartridge pouch is the only specimen known to the author. It was a trial piece, but no documentation has been found to absolutely pinpoint its period and place of trial.

This pouch served as the prototype of the next generation pistol cartridge pouch, the 1874 Pattern–a piece only slightly less rare than this 1872 Pattern.

Photo 3.

Pattern 1872 Dyer Carbine Cartridge Pouch

Photo 1.

Photo 2.

Calibre: .50/70 Government

Capacity: Estimated 40 rounds

Measurements:

OAW–	6 3/4"
OAH–	6"
OAT–	1 3/4"

Colour/Materials/Construction: Like its predecessor, the Sling Pouch "E", the 1872 Pouch was also of soft black leather. However, it does not have an inner flap, and it was designed to only be worn on the waist/sabre belt. As can be seen in Photo 2, its belt loops are exceptionally large and would fit easily over normal belt attachments.

A tool pocket on the face of this pocket (substituted for the earlier revolver cartridge pocket) is also outlined by a large semicircle of stitching, but is closed at the top (Photo 3) by a centrally-mounted button.

Like the earlier Dyer, the M1872 was also lined with white sheep's wool. Also like the Sling Pouch "E", the sides of the M1872 were gusseted, giving it the ability to slightly expand or contract with the load carried.

Markings: Large "US" embossed on the flap. Inspector's initials "J.C." stamped on tip of closing tab.

Comments: The Dyer concept first issued as the Sling Pouch "E" was refined into this new M1872 Pouch. It, also, was manufactured at Watervliet Arsenal in very limited, trial quantities as part of a M1872 (Trial) Cavalry Brace System. The brace system was very unpopular with the cavalry and was dropped after the trials, but the Dyer pouch concept was considered by many officers polled by the Ordnance Department to be a popular, adaptable and robust accoutrement. It saw further redesign and manufacture as the M1874 Dyer Carbine Cartridge Pouch.

One rarely-seen variation of the M1872 Dyer is a cut-down version, modified and issued after the adoption of the 1874 Dyer. The cut-down 1872 is seen both with and without closing tab escutcheon plate and "US" in oval embossed on the cover.

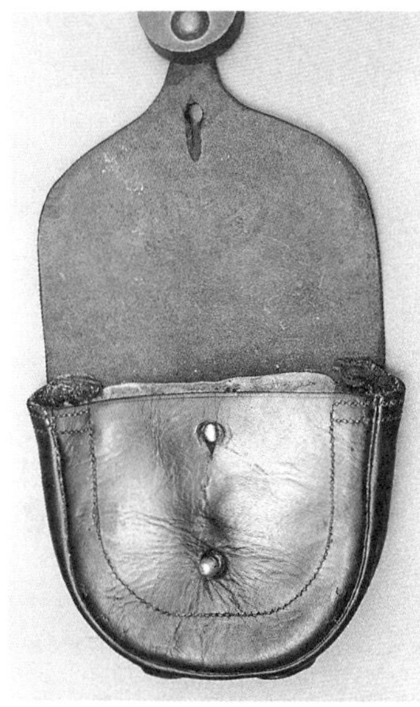

Photo 3.

Pattern 1874 Dyer Pistol Cartridge Pouch

Calibre:. 45 Pistol (.45 Colt)

Capacity: Unknown, but at least 12 rounds.

Measurements: Regulations called for a pouch of the following size:

OAH–	3 5/32"
OAW–	3 1/2"

The pouch shown measures as follows:

OAH–	3 3/4"
OAW–	3 7/8"

Colour/Materials/Construction: The 1874 Pistol Cartridge Pouch shown in Photo 1 (a drawing from Ordnance Memo No. 18 dated 1874) is the pouch design adopted. Even this drawing, however, differs from the few extant specimens known–and they differ from each other somewhat. Note that the "US" embossed on the cover of actual specimens was documented in the text of the Ordnance Memo, but was some how left off the drawing.

The pouch shown in Photos 2-4, conforms closely to the regulation pattern shown in Photo 1.

This pouch has the following features in common with its predecessor, the 1872 Dyer Pistol Cartridge Pouch:

Two belt loops
White fleece lining
Two "ears" extending over the

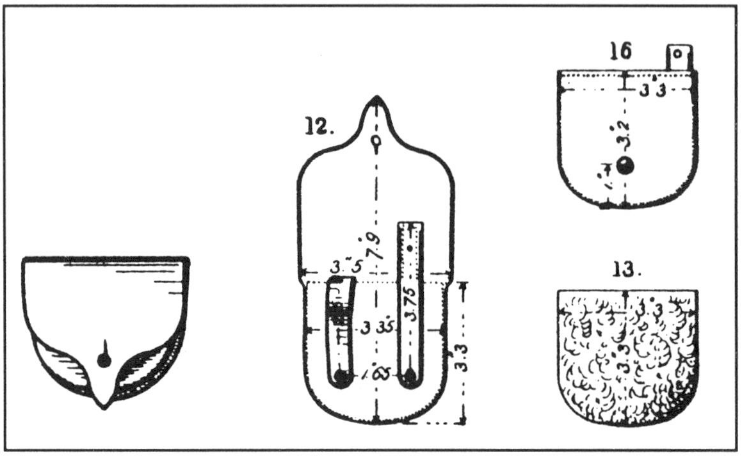

Photo 1.

Photo 2.

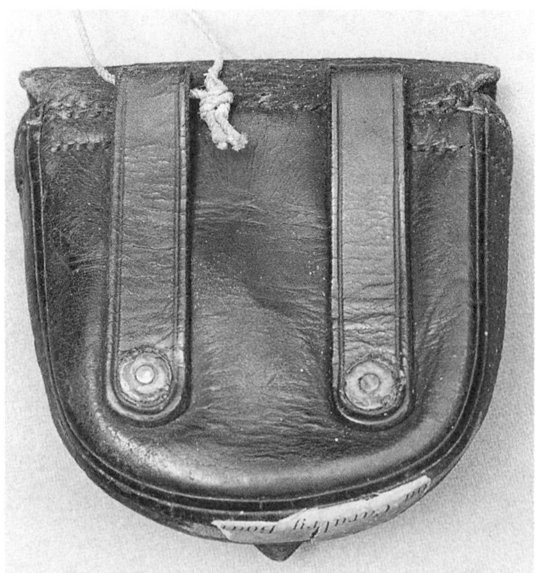

Photo 3.

Photo 4.

sides of the pouch cavity

Finial mounted low on the face of the pouch body

Construction of soft black collar leather

Note that the pistol cartridge pouch does not have a closing tab escutcheon, even though it was felt necessary for the adopted 1874 Carbine Cartridge Pouch. It would seem likely that that would have been a next design step, had the pouch actually been put into production.

Markings: A **"US"** in an oval embossed on the cover. Interestingly, that mark was not shown in the drawings from the 1874 Ord. Memo.

It is also stamped **"WATER-VLIET ARSENAL"** in two lines on the upper face of the pouch body.

Comments: The 1874 Dyer Pistol Cartridge Pouch was initially intended as a companion piece to the carbine cartridge pouch. However, though a newly-designed carbine cartridge pouch was necessary (i.e. no CW surplus pouch or box would fill the requirement), the role of the pistol pouch was easily filled by a continued issue of the CW cap pouches under General Orders 30. In short, though there were some advantages to the new pistol pouch, the economically pressed post-War Army had far better uses for their funds, and literally mountains of surplus cap pouches that could be issued as a companion piece to the 1874 Carbine Cartridge Pouch.

This pouch falls into the rare category. It is believed that all specimens were manufactured at Watervliet Arsenal.

The 1874 Dyer Carbine Cartridge Pouch-Overview

The 1872 Army trial of approximately 2,000 sets of the cavalry brace system and its attendant Second Pattern Dyer Carbine Pouch resulted in the rejection of the irksome brace system, but a large number of cavalry officers did endorse the new, belt-mounted cartridge pouch. Actually, in response to an 1874 Ordnance Department questionnaire, 26 of 43 officers polled expressed a preference for the Dyer pouch over either the Hagner No. 1 or No. 2 Cartridge Box for field service. The Hagner No. 1 was preferred, however, for garrison service. These preferences were considered by the Cavalry Equipment Board and are contained in Abstract C of Ordnance Memo 18, the proceedings of that Board.

Another abstract of that important memo, Abstract G, was a poll of 127 cavalry officers in which the scouting ("prairie") belt concept was offered as an option against both Hagner Boxes and the Second Pattern Dyer Pouch. Almost two thirds of the respondents endorsed the scouting belt for carrying ammunition in the field. This poll is particularly interesting for two reasons: First, it showed the enormous field preference for the utilitarian scouting or "prairie" belt–even though the Cavalry Board chose to recommend the design and production of the well-known Model 1874 Dyer Carbine Cartridge Pouch and the 20-round Hagner Belt Loops, rather than a true cartridge belt. The second interesting bit of data is buried in the roster of the officers polled. LTC George A. Custer and the officers who would become the commanders of the companies destined to die with him two years later all preferred the Dyer Pouch to any other method of carrying ammunition–in the field or garrison. This marked 7th Cavalry preference for the pouch leads the author to think that, possibly, the Model 1874 Dyer Pouch adopted and issued as a replacement to the aging Second Pattern Dyers still in service may have been the primary ammunition carrier for the companies who fell on Custer Hill.

Whatever the case was on that lonely hill in 1876, the 1874 Cavalry Board offered recommendations on the Second Pattern Dyer that would result in the trim and utilitarian, fleece-lined Model 1874 Pouch. Early specimens were made at Benecia, Rock Island and Watervliet Arsenals, and differed slightly in escutcheon and other, minor points. A few existing, very early specimens even lacked the brass closing tab escutcheon, but these are most unusual, transitional pieces.

By 1879, the Dyer Carbine Cartridge Pouch was officially replaced by the McKeever Cartridge Box but, in truth, by that time, field use of the prairie belt had long overtaken the Dyer.

Early, Transitional 1874 Pattern Dyer Carbine Cartridge Pouch

Comments: This pouch is exactly like other Arsenal-manufactured 1874 pouches, except that is lacks a tab closing escutcheon.

One of the criticisms made of the 1872 Dyer design was that the closing hole wore in constant use–eventually not being able to hold the flap closed. This, of course, could allow cartridges to be jolted out in normal service.

A few, very early 1874 Pattern Dyer Carbine Cartridge Pouches were obviously made, but field comments effected a design change by the time the 1874 Ordnance Memorandum No. 18 was published (showing the brass escutcheon on the regulation pattern.)

Photo 1.

First Pattern 1874 Dyer Carbine Cartridge Pouch Benecia Arsenal Manufacture

Calibre: .45/70 Government

Capacity: Listed at 40 rounds, but the pouch will hold far fewer rounds than that.

Measurements:

OAW–	6 3/4"
OAH–	5"
OAT–	1 7/8"

Colour/Materials/Construction: This soft, black leather pouch is a redesigned and reduced-size version of the 1872 Dyer Pouch. It retains the successful design elements of the 1872 Dyer: expansible gussets, belt loops, front-mounted closing finial and general shape. Photo 1 is a portion of the original Army drawing of the construction of this pouch shown in Ordnance Memo 18, dated 1874.

Where the 1874 Dyer differed was primarily in its reduced size, for the newly-adopted .45/70 cartridge. Theoretically, the pouch would hold 40 rounds, but this is not possible. The interior front and rear faces of these pouches were lined with white sheep's wool to protect the soft copper cartridge cases. As shown in Photo 4, the top of the pouch cavity had two overlapping "ears" to keep cartridges from bounding out–particularly while riding.

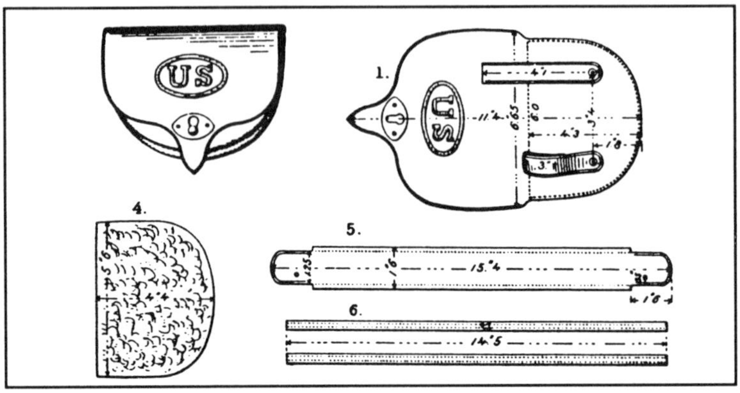

Photo 1.

The belt loops, Photo 3, were reduced in length and the lower ends raised so that the pouch would fit over the 1874 Waist/ Sabre Belt.

One key change was the development for the closing tab. Previous all-leather pouches, when empty, were difficult to button closed because the soft face of the box (holding the finial) pushed away when an at-

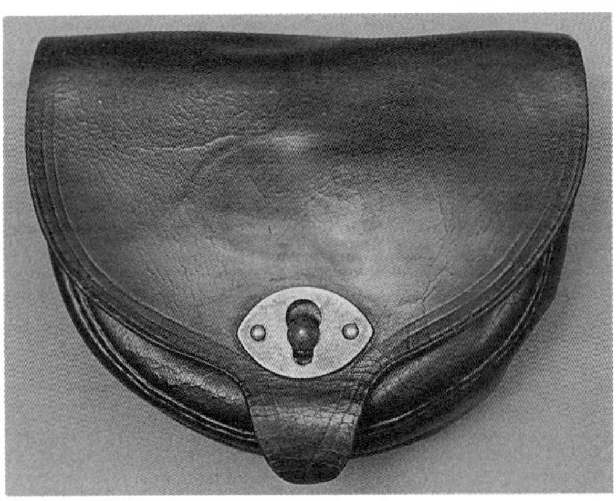

Photo 2.

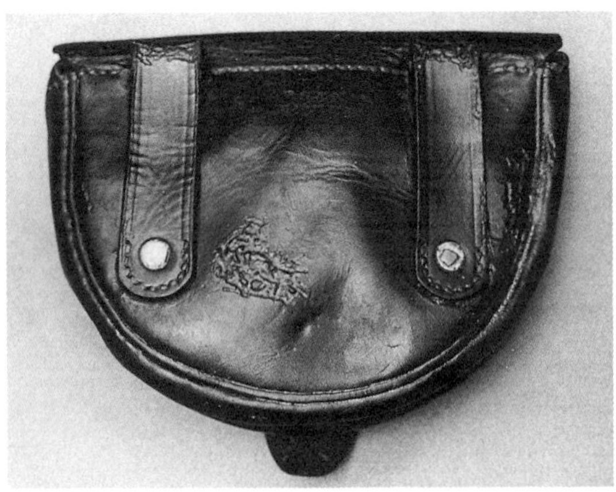

Photo 3.

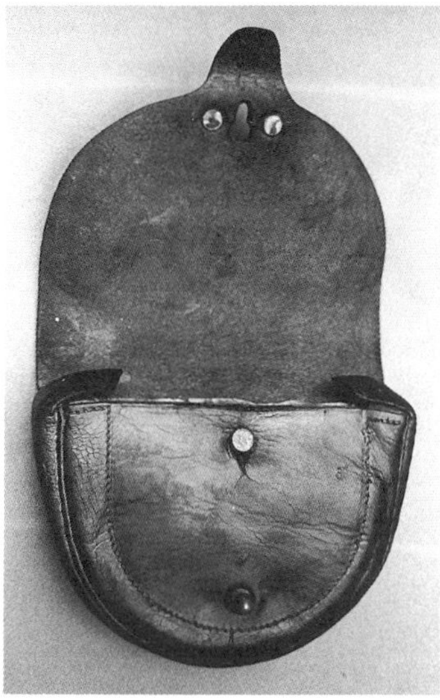

Photo 4.

tempt was made to button the tab. Also, when the hole in the tab wore, the flap would pop open. Dyer's use of a tapered hole in a brass escutcheon solved both of these problems to a large degree. The first pattern escutcheon is shown in a cut from the original 1874 Ordnance Memoranda No. 18 (Photo 1). This escutcheon had two rivets on a "cat's eye" shaped escutcheon. This early regulation shape is most often seen on the pouches made at Benecia Arsenal. Apparently, the arsenals disagreed on the optimal shape for an escutcheon plate, and even the type of closure saw experimentation–finally settling on the most commonly-seen vertical oval shaped escutcheon with four rivets. This later type is found on both Rock Island and Watervliet Arsenal-produced pouches.

An important carryover from the 1872 Pouch was the use of a face-mounted implement ("screwdriver") pouch with a button closure at its top. (Note that the tool actually carried was in reality a combination tool and could have been one of several types then in service.)

Markings: Beside the standard "US" in an oval embossed on the cover flap, this pouch is stamped:

U.S.
ARSENAL
BENECIA

on the back, between the tops of the two belt loops.

Note: The earliest Watervliet Arsenal-manufactured 1874 Dyer Pouches are also marked in three lines:

U.S.
ARSENAL
WATERVLIET

Another distinctive feature of the early Benecia Pouch is that the decorative outlining of the cover consists of three lines–as opposed to a single line found on the pouches made by the other two arsenals.

Pattern 1874 Dyer Carbine Cartridge Pouch Carpenter Turn-Key Fastener

Photo 1.

Photo 2.

Calibre:. 45/70 Government

Capacity: Listed at 40 rounds but the pouch will hold far fewer than that.

Measurements:

OAW–	6 3/4"
OAH–	5"
OAT–	1 7/8"

Colour/Materials/Construction: All features of this white sheep's wool-lined pouch are as standard–with the exception of its incorporation of its use of the experimental Carpenter Turn-Key Closure.

The brass escutcheon is attached with three flush-ground rivets. The bottom of the face of the pouch body holds a pivoting brass arm that rotates clockwise 270 degrees from the closed position (Photo 1) to a 12 o'clock position that allows the pouch flap to be lifted. The swinging arm or stud is held by friction within its base and against the escutcheon plate.

Markings: Standard **"WATERVLIET ARSENAL"** in two lines between the tops of the two belt loops on the back of the pouch.

Comments: Captain L.H. Carpenter, 10th Cavalry, was a member of the Board of Officers whose proceedings resulted in Ordnance Memoranda No. 18, covering a wide variety of cavalry items proposed and recommended for adoption. Captain Carpenter's idea addressed a real need for a more secure method of closing tabs on cartridge boxes and pouches. In this case, for the 1874 Dyer Pouch. The Carpenter Turn-Key Closure, though effective, was considerably more expensive than the basic Dyer Closure and was not deemed satisfactory for that reason. Only a few boxes were probably made for display or limited trial.

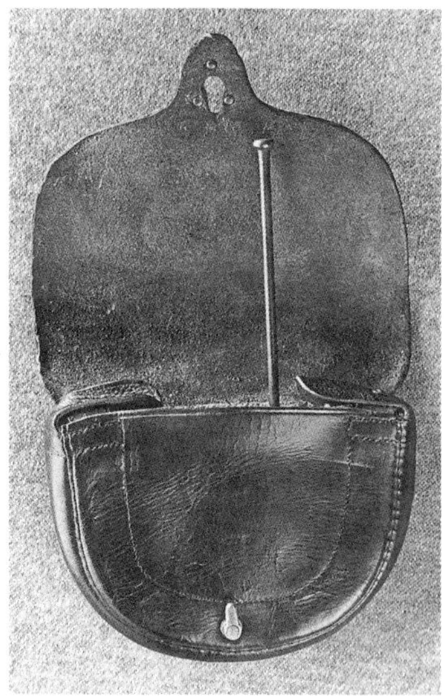

Photo 3.

Model 1874 Dyer Carbine Cartridge Pouch Wire Escutcheon Retainer Variant

Comments: A very unusual, possibly experimental M1874 Dyer Pouch. This pouch differs from the standard pattern in four ways:

1. Experimental brass-wire retention of removable tab escutcheon. The two small studs that protrude through the flap are rectangular in cross-section and the leather shows the imprint of a compression punch used to flatten the leather around the stud and thereby expose the small hole in the stud to the brass retention wire.
2. No embossed "U.S." on cover.
3. No markings of any kind, leading the owner and the author to believe this may be a unique piece.
4. Shape of the tab escutcheon is the type used on early Watervliet and Benecia Arsenal M1874 Pouches. It is believed this pouch was manufactured at Watervliet Arsenal but Benecia manufacture is suggested by the two-line edging on the face of the flap.

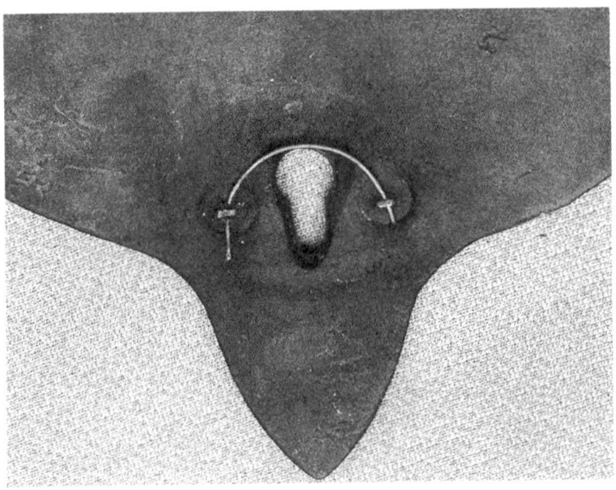

Photo 1.

Photo 2.

Pattern 1874 Dyer Carbine Cartridge Pouch Early Watervliet Arsenal Manufacture

Photo 1.

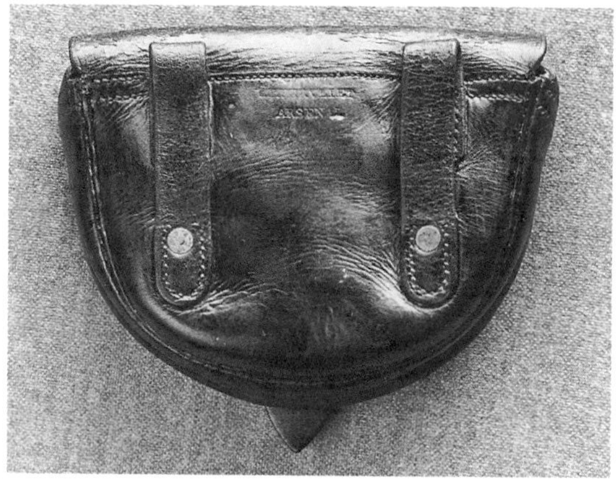

Photo 2.

Calibre: .45/70 Government

Capacity: Listed at 40 rounds but the pouch will hold far fewer than that.

Measurements:

OAW–	6 5/8"
OAH–	5"
OAT–	1 7/8"

Colour/Materials/Construction: This pouch superficially appears to be of standard pattern, but it does differ in two ways. First, it uses a three-rivet vertical brass escutcheon plate. This plate is more pointed in shape than a true oval and is distinctively of Watervliet manufacture. Second, as shown in Photo 3, the pouch has no provision for an implement pouch closing button. This may have been a trial piece to determine whether the pouch actually needed a closing button or not. (The tools issued during this period fit well down into the pocket, and really would not seem to require a special closing button.)

Markings: Standard embossed **"US"** in an oval on the cover, as well as **"WATERVLIET ARSENAL"** stamped in two lines on the back of the pouch between the tops of the two belt loops.

Comments: Considerable importance was attached by Rock Island and Watervliet Arsenals to the shape of the escutcheon plate and the number of rivets attaching it. Before both arsenals settled on the vertical oval plate attached with four rivets, Watervliet Arsenal continued to produce its own distinct design.

Note that the plate is more pointed than the plate finally settled on and shown on the Rock Island Pouch following.

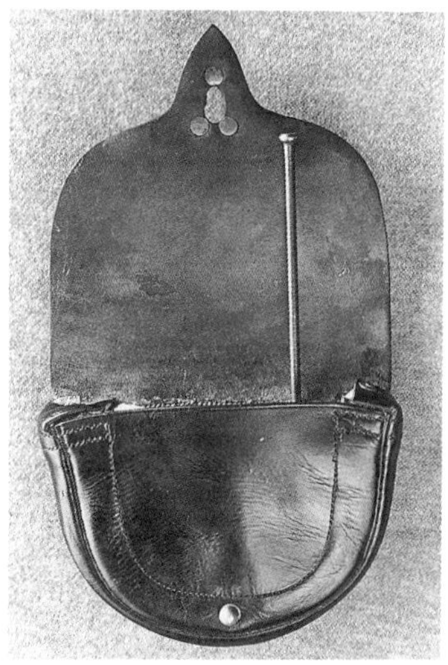

Photo 3.

Pattern 1874 Dyer Carbine Cartridge Pouch Rock Island Arsenal Manufacture

Photo 1.

Photo 2.

Calibre: .45/70 Government

Capacity: Listed at 40 rounds but the pouch will hold far fewer than that.

Measurements:

OAW–	6 3/4"
OAH–	5"
OAT–	1 7/8"

Colour/Materials/Construction: This Rock Island Arsenal Pouch is of the construction most commonly seen in 1874 Dyer Pouches. It has the late-style vertical oval, four-rivet closing escutcheon and the red sheep's wool lining that is a variant to the more commonly-seen wool lining.

Markings: This pouch has the 1874-era Rock Island Arsenal "rising sun" stamp centered on the back of the pouch, between the tops of the two loops.

The left belt loop is stamped **"D.A.M."** and the right is stamped **"W.L.F."** Both marks are inspector's initials.

Comments: This particular pouch is most often seen in very good to near new condition, but none of the Dyer pouches are common today.

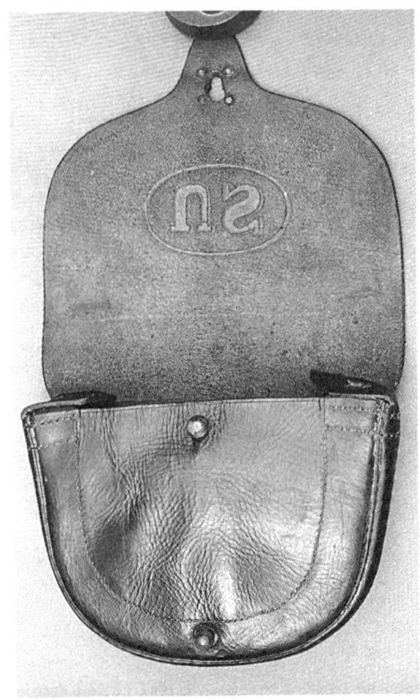

Photo 3.

Cut-Loop Variation Of M1874 Dyer Carbine Cartridge Pouch

By the time the First Pattern Model 1876 "Prairie" Belt was manufactured and issued, the M1874 Carbine Cartridge Pouch or Dyer Pouch had been widely issued throughout the cavalry. The Prairie Belt was, by far, the preferred method of carrying cartridges in the field, and many field expedient belts had been manufactured and used in cavalry units before final Army brass acquiescence to field preference for the looped cartridge belt. So, in 1876, the Dyer Pouch and the Prairie Belt were both in use.

The First Pattern M1876 "Prairie" Belt, at 2 inches, was narrower by 1/2 inch than subsequent patterns, allowing the lead bullet tips to protrude and mark the soldiers' uniforms. There was one saving grace, however. The earliest pattern M1876 was so narrow that, by cutting the loops on the Dyer Pouch, the pouch could be thonged onto the prairie belt (Photo 1), thereby adding at least 20 more important rounds of carbine ammunition to the cavalryman's field gear. Since the loops of the pouch interfered with only two or three loops on the prairie belt, the net gain to the trooper was a 50% increase in easily-available ammunition. Further, the pouch could be positioned or re-positioned anywhere on the belt for comfort, balance or ready availability.

Photo 1.

Examination of several specimens of this early modification to the Dyer Pouch shows a high level of workmanship and consistency of manufacture, arguing for modification at a depot or arsenal. The surviving number of these pouches also suggests that a number of them were done.

The work, as shown in Photo 1, consists of a cutting of the belt loops at midpoint, then punching a small, neat hole near each of the four cut ends. Sturdy leather thongs were looped through the punched holes and the pouch was ready to be mounted onto a leather sabre belt, or onto the early, narrow cartridge or prairie belt.

When field complaints about the bullets marking soldiers' uniforms reached Watervliet Arsenal, the next manufacture of the M1876 "Prairie" belts made them 1/2 inch wider–solving the marking problems, but making the belts too wide to mount the modified Dyer Pouch. This indicates that the actual service period of the modified Dyer was early in the Indian War, but quite short-lived.

Photo 2 shows a front view of the modified Dyer Pouch mounted on a First Pattern M1876 Cartridge or "Prairie" Belt.

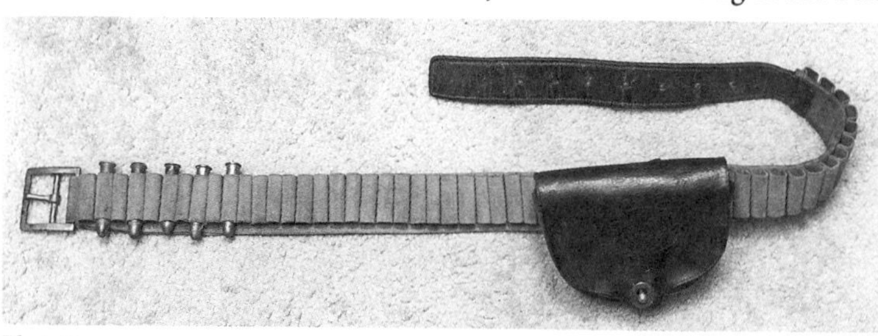

Photo 2.

The Hagner Cartridge Boxes

Overview

In 1870, Colonel P.V. Hagner, Commanding Officer of Watervliet Arsenal, submitted two patterns ("pairs") of cartridge boxes to the 1870 St. Louis Ordnance Board for consideration. Hagner's cartridge boxes (submitted as a set of two) were part of a grouping of 12 different designs submitted to the board and were given the alphabetic designation "M". Fifty pairs of Hagner's 1870 Trial Boxes were manufactured.

The Hagner Boxes were sufficiently satisfactory that they were only slightly modified in design and were manufactured at Watervliet Arsenal in some quantity for the infantry and cavalry field trials directed by the Army Chief of Ordnance in 1872. The taller, flatter Hagner Box was stamped **"No. 1"** and the lower and thicker Hagner Box was stamped **"No. 2."**

The trials each tested a different brace system. The Hagner Boxes were in .50 calibre for the then-current rifles and carbines, and were identical between the infantry and cavalry field trials.

The 1872 Infantry Brace System consisted of the following:

1 Waist Belt

1 Bayonet Scabbard, steel, with swivel frog

2 Cartridge Boxes, to hold each 24 Metallic Cartridges, carried in loops (of two patterns for trial)

1 Valise to hold clothing, with pouch under flap to receive two packages extra cartridges (40)

1 Brace Yoke

2 Coat Straps

As regards the cartridge boxes, the 1872 infantry manual stated, "While the object is to provide means of carrying a good supply of ammunition–it is proposed to carry only one Cartridge Box on ordinary occasions, in peace–." Photo 1 shows two views of this system, but does not well distinguish the greatly different shapes of the two different cartridge boxes. The manual goes on, "During drills, or on garrison duty,–it is proposed that only one Cartridge Box holding 24 rounds be carried. This is sufficiently supported by the Waist belt and may be placed either in front or behind." This latter carriage is shown in Photo 2.

The 1872 Cavalry Brace System field trial pitted the two Hagner pattern boxes against the Second Pattern Dyer Pouch for field and garrison use.

The results of the infantry field trial were not supportive of either of the Hagner Boxes. The infantry would continue to look for a more satisfactory cartridge box and, in the end, find the McKeever Cartridge Box satisfactory for garrison use and the increasingly popular "scouting" or "hunting" belt for field service. But, this result would take a few years more.

The results of the 1872 cavalry field trial were somewhat more interesting. Ordnance Memo. 18 contains at least two important polls of cavalry officers whose troops tested various concept boxes and the Dyer Pouch against the conceptual scouting or "prairie" belt. In a test of Hagner Boxes against the Second Pattern Dyer Pouch, the pouch and belt were preferred for field service. For garrison service, the Hagner No. 1 was almost three times as popular as the more compact No. 2.

Both of the Hagner pattern

boxes had a problem with the mounted service–the cartridges tended to bounce out of their loops over time. In addition, when the Hagner No. 2 was worn at the small of the back, the rubbing of the cantle released the closing tab from the bottom-mounted finial. This was solved, for the short period of its field use, by a reversal of the finial and closing tab (shown later in this section.)

When the calibre of Army weapons changed to .45, the Hagners changed also–with Rock Island Arsenal-applied modifications to each model. However, by 1874, the adoption of the new cartridge and the favorable test of the McKeever Cartridge Box, the Hagner Boxes were being retired completely from the Regular Army. Since the state militias were armed with .50 calibre rifles and carbines, the Hagner Boxes were easily assimilated. Later, as the state militias were armed with .45 calibre longarms, Hagner Boxes were modified at Rock Island Arsenal for continued militia service. The leather strip modification of the Hagner appears in an Ordnance Office letter to the Chief of Ordnance, Military Division of the Missouri, dated 22 May 1876:

"Sir: I send you by mail to-day a caliber .50 cartridge box, adapted to the cal. .45 cartridges by inserting leather strips which project into the loops...The change in the interior of the box MAY necessitate the lengthening of the button hole of the billet a quarter of an inch or so.
Regards
S.V. Benet
Brigadier General,
Chief of Ordnance"

Another correspondence relating to this modification process was a 29 May 1876 Special Order No. 99 from the HQ to the Military Division of the Atlantic. Paragraph 2 from that S.O. reads as follows,

"Commanders of such companies of Artillery and Infantry in this Division as have on hand SERVICEABLE cartridge boxes Nos. 1 and 2, caliber .50, will send them without delay, to Watervliet Arsenal...with a view to their adaption to the caliber .45 ammunition, by insertion of leather strips projecting into the loops . . ."

Similar orders probably went out to all other military divisions for a total turn-in of the Hagner Boxes. It is not felt that the modified boxes were returned to Regular Army service, but remained in inventory for issue to state militias, upon their request.

There are at least three modifications to the Hagner Boxes:

1. The standard .50 calibre boxes were converted to .45 calibre to serve the new .45/70 round.
2. The Hagner No. 2 Box was sometimes modified from rifle or carbine cartridge use to a shotgun-like salute gun cartridge.
3. The Hagner No. 2 Box's finial/tab arrangement was sometimes modified for better cavalry service (in .45 calibre).

The calibre modifications of the Hagner No. 1 Box was driven by its height. The cartridges of the box were held tightly by the stiffened canvas loops stitched to the leather box walls. When the calibre was reduced from .50 to .45, all three rows of loops had to be reduced in dimension, since the lower two rows of cartridges did not bear against a top flap (which would have served to keep them in place.) Simply, the loops had to do all of the work of cartridge restraint. To accomplish this, a light leather strip with eight downward-projecting fingers was inserted over and down into each row of eight loops. The fingers project 3/4" down into the back of each cartridge loop, thereby reducing the inside diameter to a snug fit for a .45/70 cartridge case. These leather strips were then stitched at the top center and each end of the row of eight loops.

The calibre modification for the Hagner No. 2 was even simpler. Apparently, the Ordnance designers felt that only the outer (face) row of cartridges was at risk with the new calibre. So a simple 3/8" wide strip of leather, 5 3/4" long was stitched to the exterior top lip of the cartridge box body–just at the top of the loops on the face of the box. This made an interference fit for the .45 calibre cartridges being inserted (tipping them slightly forward) and cartridge retention was deemed adequate. Since the interior 16 rounds sat deep into the box body and were somewhat restrained by the cover of the box, no additional loop size reduction was felt necessary. (Remember that the difference in loop and case size is, practically, quite minor.) Howerver, the length of the newly-adopted .45/70 round did develop one problem for the Hagner No. 2. When the back row of the box was filled with cartridges, the box could not be closed without stretching or enlarging the hole in the closing tab and many boxes are seen today showing that stress.

The Hagner No. 2 modification to salute gun cartridge was effected by a simple removal of the cartridge loops and the insertion of a soft wood block bored for six salute gun cartridges. These cartridges, actually blank shotshells, were used in sub-calibre devices mounted in breech-loading cannon used for ceremonial functions.

The modification of Hagner No. 2 Boxes for more efficient cavalry use is more interesting. The cavalry appeared to carry the No. 2 box on the sabre belt at the small of the back. During riding, the rubbing of the cantle of the saddle tended to push the cover-mounted closing tab open, allowing cartridges to bounce free. To remedy this, some Hagner No. 2 Boxes were modified by reversing the positions of the closing tab and the finial. That is, the finial was moved up onto the flap and the tab secured on the bottom of the box (as shown later in this section). The net effect was that the closing tab had to pull out and down to open.

The Hagner No. 1 Box had one built-in human engineering flaw. Its height was somewhat awkward when worn to the front as prescribed by regulation, i.e. it prevented the wearer from bending far forward. Colonel Hagner must have been aware of this deficiency, because he designed a V-shaped restraining device on the back of the box. This device resisted the bending of the box (and consequent breaking of stitching) occasioned by the wearer's bending forward against the top of the box. The more conventional and lower design of the Hagner No. 2 Box required no such restraint and did not discomfit the wearer unduly. Due to its height and complaints of cartridge loss (probably while riding), the Hagner No. 1 was not well accepted by the cavalry for mounted use.

Photo 1.

Photo 2.

Hagner Number 1 Cartridge Box

Calibre: .50/70 Government

Capacity: 24 rounds in three rows of eight cartridges each. Two lower rows; one upper row.

Measurements:

OAH Flap–	5 3/8"
OAW Flap–	6 13/16"
OAT Bottom–	2"

Colour/Materials/Construction: This black leather box has standard brass fittings and rivets. The 1" cartridge loops are made of stiffened, light-coloured, heavy canvas loops, stitched onto 1 1/2"-wide backing strips of webbing in three rows of eight loops each.

The body of the cartridge box is formed of three main pieces of leather: The central portion, which serves as the flap, the top and back of the box and the face of the box body. The other two pieces are the tall, rounded and tapered end pieces. The right end of the inner box body was bulged to form a tool pocket, and the inner face of that bulge provided with a short strap, which kept the gun tool firmly in place. Gun tools appropriate to this pocket include either of the M1866 Combination Tools or the M1868, M1870 or M1876 Combination Tools. The M1879 Tool is not appropriate.

The Hagner No. 1 is noteworthy for its height–which necessitated a rather complex V-shaped restraining strap on the back of the Pouch. Since the box was so tall, the soldier's bending forward tended to bend the box and break its stitching. The V-strap was then incorporated to resist this bending of the box and, in effect, operate as a stiffener.

Markings: A **"U.S."** in an oval is center-embossed on the front of the flap. Just between the tops of the two belt loops and just above the cartridge loop stitching on the exterior of the box, **"No. 1"** is stamped. This particular speci-

Photo 1.

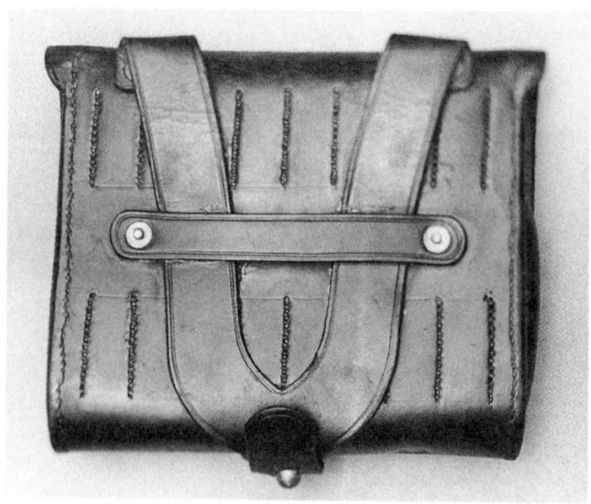

Photo 2.

men bears the inspector's stamp, **"J.C."** at the end of the closing tab.

Comments: The No. 1 Box was not popular outside of garrison and most specimens seen today are in very good to new condition. They are, however, quite desirable, especially in original .50 calibre.

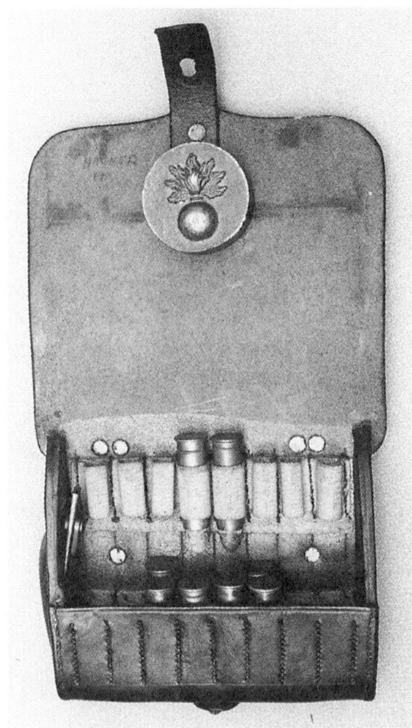

Photo 3.

Modified Hagner Number 1 Cartridge Box

Calibre: .45/70 Government

Capacity: 24 rounds in three rows of eight cartridges each. Two lower rows; one upper row.

Measurements:

OAH Flap–	5 3/4"
OAW Flap–	5 3/4"
OAT Bottom–	2"

Colour/Materials/Construction: A black leather box with standard brass fittings and rivets. The 1" cartridge loops and the backing to which they are sewn are stiffened canvas.

The conversion of this standard .50 calibre box to .45 calibre was cleverly done by the leather strip modification method, described earlier in this section. In Photo 2, a pointer indicates the leather strip applied over the cartridge loops and with leather "fingers" extending down the backs of the loops. The leather strip covers almost all of the two brass rivets shown.

The arrows in Photo 3 show the points at which the leather calibre conversion strips were sewn into the box.

Markings: A **"U.S."** in an oval is center-embossed on the front of the flap. Just between the tops of the two belt loops, and just above the cartridge loop stitching on the exterior of the box, is stamped **"No. 1."** This particular specimen is inspector-stamped on the tip of the closing tab, **"C.T.S."**

Comments: This No. 1 Box has a canvas backing to the cartridge loops. This differs from the unmodified .50 calibre box which has a web backing strip. This may indicate at least two production runs of this box, or just an in-production change of materials.

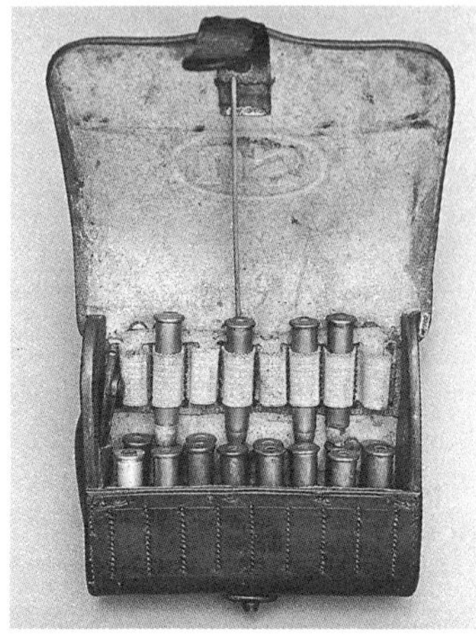

Photo 1.

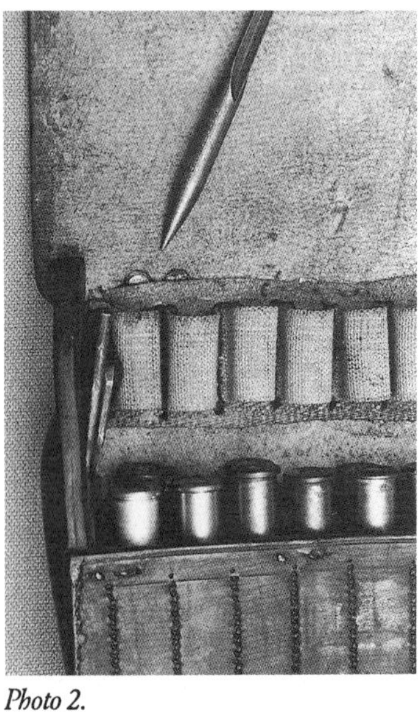

Photo 2.

Photo 3.

Hagner No. 2 Cartridge Box

Photo 1.

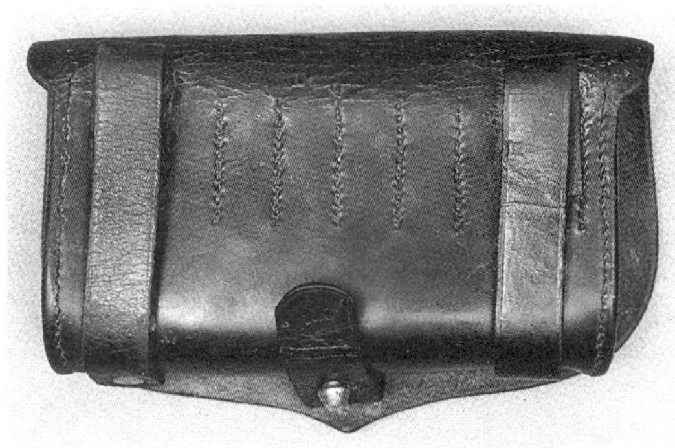

Photo 2.

Calibre: .50/70 Government

Capacity: 24 rounds in three rows of eight cartridges each. One row on the exterior of the box body face and two internal rows.

Measurements:

OAH Flap without Finial–	3 11/16"
OAW Flap–	7 1/4"

Colour/Materials/Construction: This black leather box has standard, arsenal-type brass rivets and bottom finial.

The 1 1/4" stiffened canvas loops are mounted on 1 1/2" strips of tan webbing. The two front rows of cartridge loops are stitched, back-to-back, through the front face of the box body. The back row of loops is 3/4" higher than the other two rows. Like all true Hagner No. 2 Boxes, the box body and flap are all one piece of leather, with only the rounded end caps and loops added.

As standard with all Hagner No. 2 Boxes, the right wall of the box body is bulged outward to accept any of the combination tools from M1866 to M1876.

Markings: Centered on the cover of this box is a "U.S." in an oval. Centered on the back of the box, just above the tops of the loops, "No. 2" is stamped. An illegible set of three inspector's initials are stamped at the end of the closing tab.

Comments: The No. 2 Boxes are almost always seen showing more use than the No. 1 Boxes. The original .50 calibre specimens are more desirable than modified No. 2 Boxes.

Photo 3.

Modified Hagner Number 2 Cartridge Box

Calibre: .45/70 Government

Capacity: 24 rounds in three rows of eight cartridges each. One row on the face of the box body and two internal rows.

Measurements:

OAH Flap without Finial–	3 11/16"
OAW Flap–	7 1/4"

Colour/Material/Construction: A standard Watervliet Arsenal Hagner No. 2, converted with the small exterior strip of leather above the row outside loops. Photo 2 shows a closeup of a .50 calibre, unmodified box face. Compare that with Photo 3, showing the simple modifications of the strip of leather added to the box face (at the pointer). In this specimen, the cartridge loops are web, not canvas, and have two parallel lines of dark blue stitched through their center, lengthwise. The backing material is identical webbing.

Markings: Standard

Comments: It is not known if the use of web loops indicates a follow-on production of this box or a mid-production change of materials.

Photo 1.

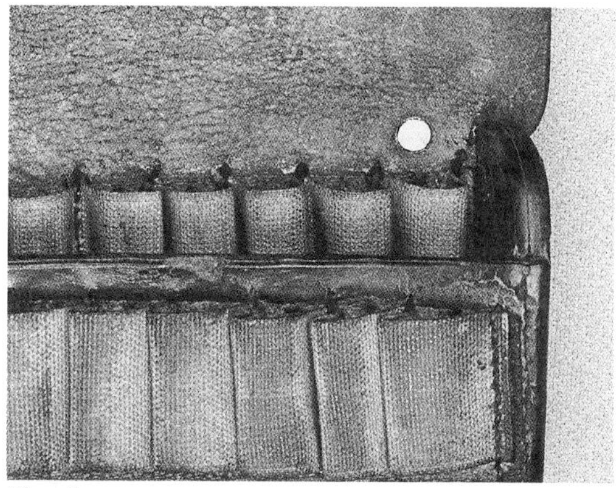

Photo 2.

Photo 3.

Modified Hagner Number 2 Cartridge Box–Cavalry Variations

Calibre: .45/70 Government

Capacity: 24 rounds in three rows of eight cartridges each. One row on the face of the box body and two internal rows.

Measurements:
OAH Flap– 3 11/16"
OAW Flap– 7 1/4"

Colour/Material/Construction: A standard Hagner No. 2 Box, converted to .45/70 Government. Stiffened canvas loops with web backing strips. The real distinction of this box is its reversal of the positions of the closing tab and finial. The finial on this box has been removed from the box bottom and placed at the top of the box flap in the hole originally occupied by the closing tab rivet. The finial is backed (inside) by a blued iron burr. The closing tab has been attached to the bottom of the box by two rivets.

Photo 1.

Markings: Closing tab has two sets of inspectors' initials: **"A.R.S."** (A.R. Smith), over **"N.M."** (an unknown inspector). All other markings are standard to the Hagner No. 2.

Photo 2.

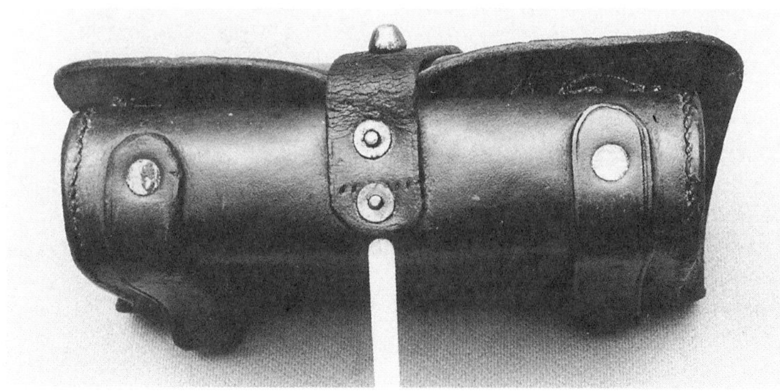

Photo 3.

Modified Hagner No. 2 Cartridge Box–Unmarked Variant

Calibre: .45/70 Government

Capacity: 24 rounds in three rows of eight cartridges each. One row on the face of the box body–and two internal rows.

Measurements:

OAH Flap without Finial– 3 11/16"
OAW Flap– 7 1/4"

Colour/Materials/Construction: This black box is a very practical conversion to .45 calibre from .50 calibre. A 3/8" wide by 5 3/4" long strip of leather has been stitched above the loops on the front face. This pushes the .45/70 cases out against the loops fairly tightly and makes them secure. However, since the top of the flap of the box bears against the distinctly taller .45/70 round, it is obvious that the arsenal re-designers (to .45 cal.) felt further loop size reduction unnecessary.

Markings: The most obvious feature of this box is its lack of a "U.S." and oval on the cover. Interestingly, this lack of marking persisted from original manufacture to conversion and re-issue. However, the most important identifying mark on this box is the Watervliet Arsenal inspector's initials on the closing tab tip: "C.G.H." This box lacks any vestige of the "No. 2" stamp almost always seen on the upper back of Hagner No. 2's.

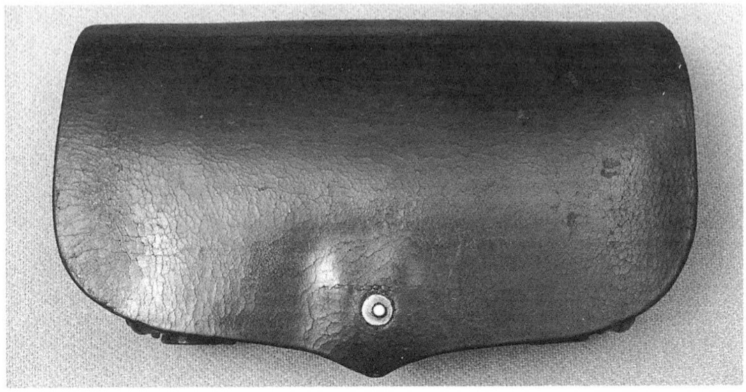

Photo 1.

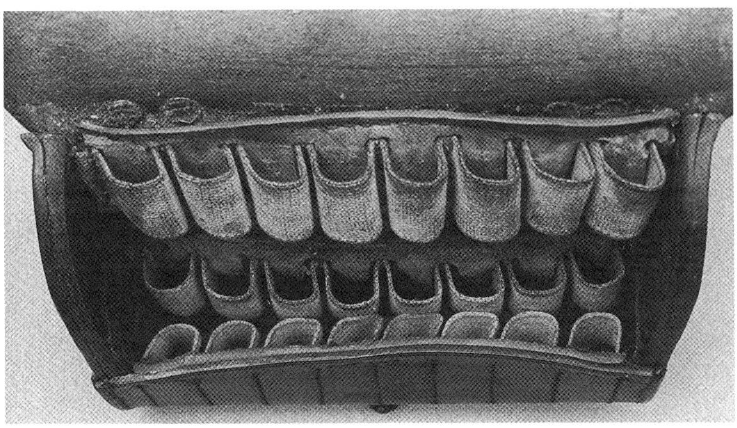

Photo 2.

Comments: An intriguing box, converted exactly as others shown, but definitely out of the ordinary.

Modified Hagner Number 2 Cartridge Box Signal Cartridge Variant

Photo 1.

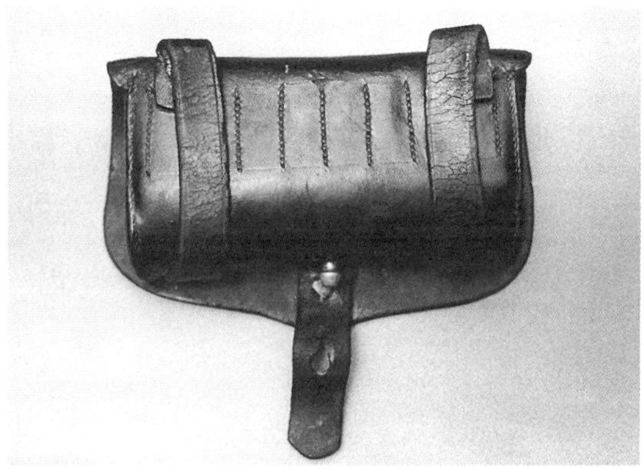

Photo 2.

Photo 3.

Calibre: 8 Gauge Signal or Salute Cartridges

Capacity: Four rounds of 8 ga. signal rounds in four holes bored in a soft wooden block.

Measurements:

OAL Flap–	3 11/16"
OAW Flap–	7 1/4"

Colour/Material/Construction: A standard Hagner No. 2 Cartridge Box, converted by the removal of all cartridge loops. The interior cavity was then filled with a bored wooden block containing four holes for 8 gauge signal or salute cartridges. These cartridges were fired in artillery sub-calibre firing devices for salute, parade, drill or demonstration.

Markings: Standard Hagner No. 2 markings. **"T.M.L."** inspector's initials on tip of closing tab.

Comments: The Hagner No. 2 has also been used for a six-round cartridge block, containing 12 gauge signal or salute cartridges. Such a use extended the life of the No. 2 Box much later than many other cartridge boxes, and it is believed to have been used for artillery salute purposes well into the 1890s, both in the Regular Army and the state militias.

State Militia Hagner No. 2-Type Cartridge Box

Calibre: 50/70 Government

Capacity: 16 rounds carried in two inside rows of canvas loops.

Measurements:

OAL Flap–	6 7/8"
OAW Flap without Finial	3 7/16"

Colour/Materials/Construction: This black leather box is very reminiscent of the Hagner No. 2 Box, but it differs in three major aspects:

1. No exterior row of cartridge loops
2. Multiple-piece construction
3. Use of commercial iron "flower" rivets

In particular, the flap and top piece is separate and attached to the back of the body section by seven black-pointed iron rivets. Similarly, the closing tab is attached with two iron rivets. The 19/32" wide belt loops are unedged, and are secured at the bottom with blackened iron rivets. At the top, they are held captive by the same rivets as affix the cover to the body.

The right end of the box body is bulged, to accept a gun maintenance tool, secured inside by a small strap. Such a tool could be either of the M1866 Combination Tools or the M1865, M1870 or M1876 Combination Tools.

For some reason, a perforated leather strip is riveted lengthwise in the interior bottom of the box–probably to serve as a stop for the bullet.

Markings: A **"U.S."** in an oval is embossed on the flap front. In this particular specimen, this oval is pierced in two places for what must have been a brass emblem or plate, also probably oval.

Comments: This may have been a commercial venture, directed toward sales to several state militias. It is not a commonly-seen box.

Photo 1.

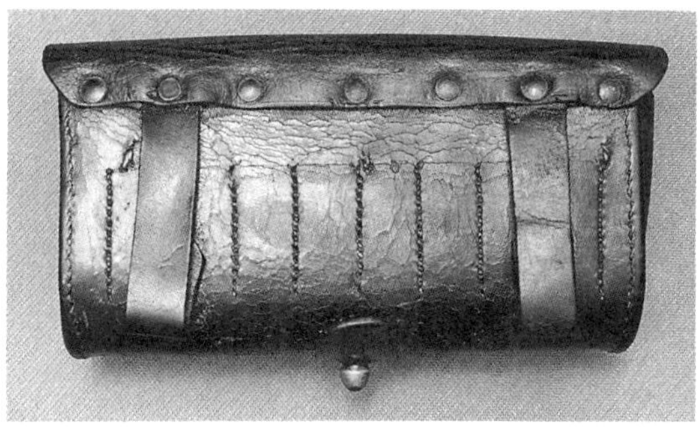

Photo 2.

Photo 3.

Rhode Island Militia Hagner No. 2 Type Cartridge Box

Calibre: .45/70 Government

Capacity: 20 rounds in two rows of ten holes each.

Measurements:

OAL Flap– 7 3/16"
OAH Flap– 3 23/32"

Colour/Materials/Construction: This black leather box was arsenal-made and incorporates a light, stepped wooden block, secured to the leather body by two brass screws through the back of the box and one, centered, through the front.

Though resembling the Hagner No. 2 Box, there is no detail similarity. The box features a single piece construction, where one piece of leather serves as the flap, back and front of the box. Two 3/4" wide loops are stitched at the bottom and iron riveted at the top.

The capital letters "R.I.M." in a 2 11/16" x 1 9/16" oval are embossed on the front flap. On the upper back of the box is stamped:

WATERVLIET
ARSENAL

Comments: A desirable militia box, probably dating from the mid to late 1870s. No state or federal variations are known.

Photo 1.

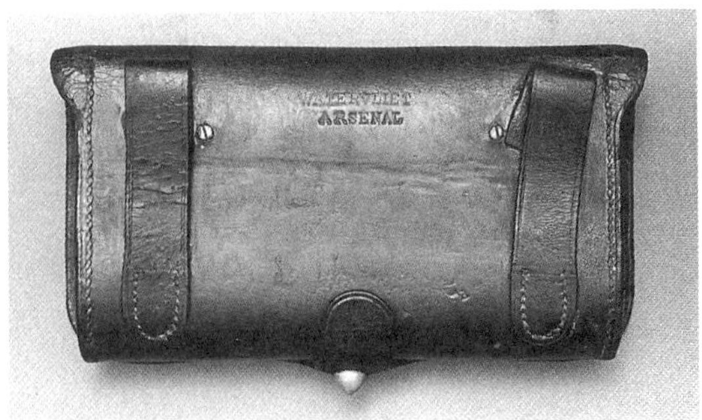

Photo 2.

Photo 3.

N.G.P. (National Guard of Pennsylvania) Hagner No. 2-Type Cartridge Box

Calibre: .50/70 Government

Capacity: 16 rounds in two rows of 8 rounds each

Measurements:

OAL–	6 3/4"
OAW–	3 1/2"
OAT–	1 3/4"

Colour/Materials/Construction: An unusual state militia box that has a superficial resemblance to the Hagner No. 2 Box. However, this box was designed around a stepped wooden block, bored with two rows of holes. The block was secured in the box body by two screws in the back and one in the front.

The box body itself is reinforced with two angled brass pieces, riveted onto the upper sides and front corners of the box.

There is no provision for an implement pocket tool pouch.

The cover of the box hinges on a covered, captive brass rod, secured just above the tops of the two belt loops.

The workmanship is of high quality and, though appears unmarked, it may have been made at Watervliet Arsenal to Pennsylvania state specifications.

Markings: None, except the **"NPG"** in an oval on the cover.

Comments: Probably Watervliet Arsenal-manufactured. The design may have been a Pennsylvania Militia idea ("favorite son"), but was no improvement over the Hagner No. 2.

Photo 1.

Photo 2.

Photo 3.

Unknown Hagner No. 1-Type Box

Calibre: Unknown

Capacity: Unknown

Measurements:

OAH–	6"
OAW–	6"
Max. Thickness Top–	2 1/8"
Max. Thickness Bottom–	1 1/2"

Colour/Materials/Construction: In construction, this unusual black leather box–shown on the right in Photos 1 and 2–has a slight, visual similarity to the Hagner No. 1. However, it differs markedly from the Hagner Box in a number of ways. It has no cartridge loops or any other fixtures that might give a clue to its purpose, e.g. carrying a cartridge box or loose cartridges, etc. It has, as shown in Photo 4, a wedge shape and, unlike the Hagner No. 1, has a high box face that gives the box a deep, wedge-shaped cavity. This cavity (Photo 3) accepts no combination of cartridge boxes of the era, and it is certain not to have been designed for loose cartridges. The cavity shows no signs of having had a fleece liner, such as that on the Dyer Pouches. A removable cartridge block is a possibility, but the wedge shape argues against a reversible block and, for one row of cartridges, the box is needlessly tall. For an unknown reason, the upper edges of the box are purposely swelled outward.

The face of the box has a centrally-mounted brass finial. The flap of the box has an embossed "US" in an oval.

The back of the box appears to be a combination belt loop and Hagner No. 1-type stiffener. (Photos 2 and 4). Most interestingly, the center section of the loop/stiffener is reinforced with riveted steel plate on the underside. Whether this plate was primarily intended as a wear protector or had some more sophisticated purpose (e.g. attachment to a fitting) is unknown.

Markings: Though rubbed somewhat flat, and therefore indistinct, is a **"US"** in an oval–of the exact size as found on the 1874 Dyer Pouch. The central, rear strap in the brace or belt loop arrangement on the back is marked **"ROCK ISLAND ARSENAL E.H.S."** The latter initials are those of an arsenal inspector. Sub-inspector initials **"J.L."** are found on the back of the box body, under the strap arrangement.

Comments: A true enigma, the wear marks on this box give no clue to its specific use and, as far as the author knows, there are no other known specimens.

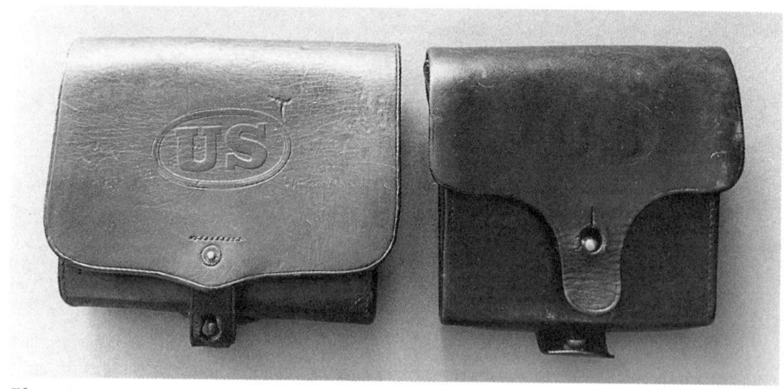

Photo 1.

Photo 2.

Photo 3.

Photo 4.

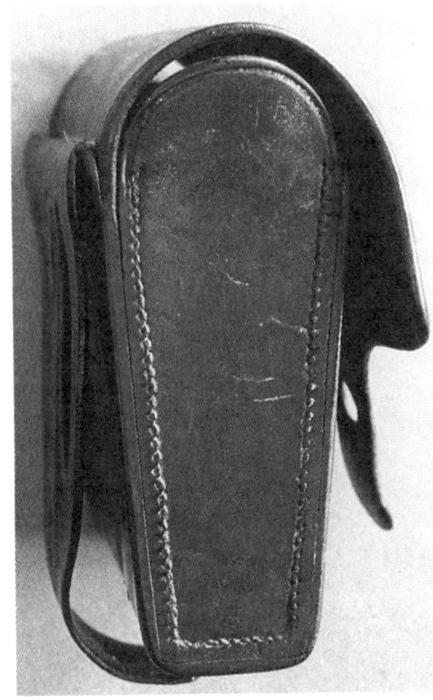

Photo 5.

Hazen's Moveable Cartridge Loops

Calibre: .45/70 Government

Capacity: 20 rounds; one row of 12 and one row of eight.

Measurements:

OAL*– 8 1/4"
OAH–with Belt Loops– 2"
OAT Loops Only– 1 1/8"

(* Note: The original specifications reflect an overall length of only 7.7 inches.)

Colour/Materials/Construction: The actual cartridge loops are made of black bridle leather and the belt loops are made of black collar leather. The entire device has a slight flex to it, so that it would curve to the wearer's waist.

The two end belt loops are 5/8" wide, while the center belt loop is 1" wide. The center loop is marked with the arsenal of manufacture.

Photo 1 shows the original drawings of the cartridge loops from the 1874 (Cavalry) Ordnance Memorandum No. 18. Photo 4 shows the back of a set of loops mounted on the standard cavalry sabre belt. This side would be against the wearer's waist. Photo 5 is a view from the top of a set of loops.

Markings: The center belt loop of each set of cartridge loops is stamped with the name of the manufacturing arsenal–either **"WATERVLIET ARSENAL"** OR **"U.S. ARSENAL BENECIA"**.

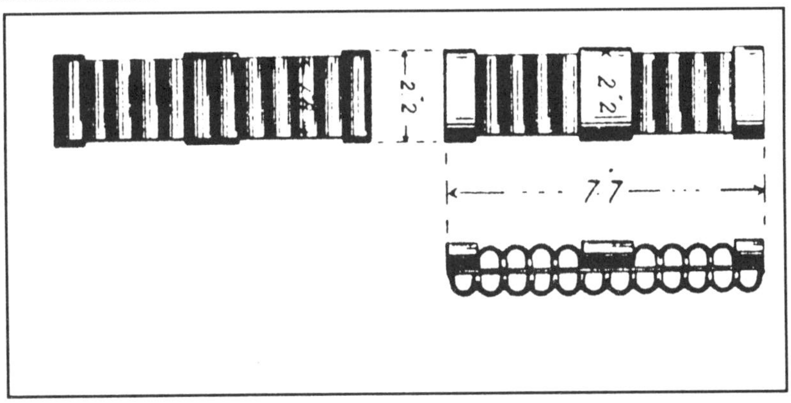

Photo 1.

Photo 2.

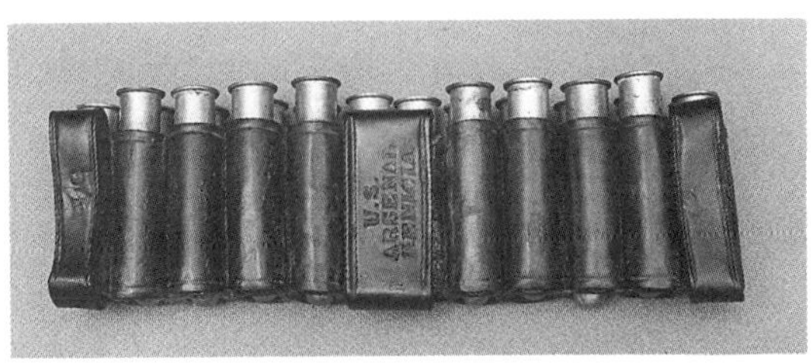

Photo 3.

Comments: By 1874, a great number of improvised "scouting" or "thimble" belts were being used by infantry and cavalry troops on the Frontier. The Ordnance Dept. was not unaware of this use, but was slow to react to the obvious preference by field troops. However, when Col. Hazen, 6th Infantry, submitted his sample cartridge loops to the 1874 Cavalry Board, the Board's poll of cavalry officers in the field indicated widespread popularity of the "scouting belt" for field use. Unwilling to make such a creative leap from conventional equipment, the Board recommended Col. Hazen's Loops as a solution. The loops had the advantage of being carried on the sabre belt only as necessary, just as the recommended Dyer Carbine Cartridge Pouch. Unfortunately, the Hazen Loops had one extremely serious drawback that trial issue would have revealed: the interface of the copper cartridge cases against the leather (containing tannic acid) provoked the formation of verdigris almost immediately. This thick, sticky, soap-like green material could seize the cartridge in the loop or cause a failure to chamber or extract from the carbine. The field-expedient scouting belts made of cloth or with cloth, posed no such problems.

Though clearly identified in the 1874 Ordnance memo as regulation, the Hazen's Loops are quite scarce. Collectors, for years, have intuitively known that the number of Loops surviving did not support the indicated basis of issue of two per cavalry trooper. And, most of the surviving specimens are in excellent condition. A National Archives letter dated 27 November 1876 from the Chief of Ordnance, S.V. Benet, to the Quartermaster General, Montgomery C. Meigs, indicates not only Benet's ignorance of the loops actually produced (at Watervliet and Benecia Arsenals) but their intended, regulation issue and use. To quote, "the leather cartridge loops referred to, must have been improvised by the soldiers, as none, as far as I know, have ever been made and issued by this Department." What that suggests is that the officially adopted Hazen's Loops were quickly recognized as a liability and, like the 1874 Dyer-pattern Pistol Cartridge Pouch, were not put into full production. It is the author's belief that production was started at the two mentioned arsenals but just as quickly halted and the Loops withdrawn from service shortly thereafter. This belief is supported by the 1875 Ordnance order that a trial of the "prairie" or "scouting" belt be done, followed by the Army's adoption of a regulation cartridge belt in 1876.

Today, the Hazen's Loops are quite scarce. Though not thought to have been used, they are an important "bridge" between the leather accoutements of the early Indian War and the more practical cloth "prairie" or cartridge belts that became predominant for the next 25 years.

Photo 4.

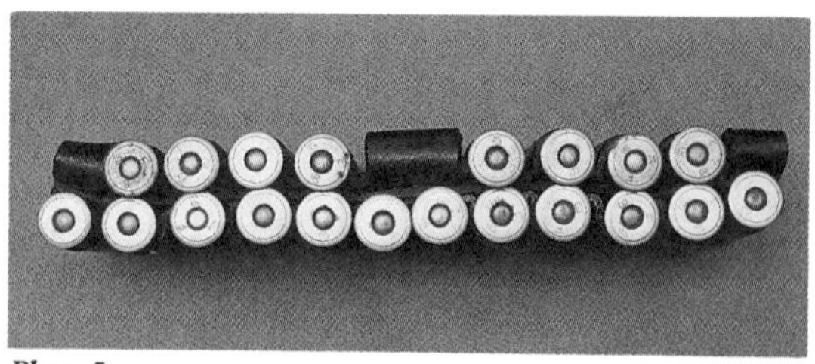
Photo 5.

The McKeever Cartridge Boxes

Overview

The Patents

On 10 June 1873, Patent Number 139,847 was granted to an infantry captain stationed in the deep South. This patent, quoted below, covered what became a popular, compact and long-lived cartridge box that was still in production in 1906. Referring to the patent drawing shown in Photo 1, the patent reads as follows, *"Be it known that I, Captain Samuel McKeever, of the United States Army, stationed at Mobile, in the county of Mobile and State of Alabama, have invented a new and Improved Cartridge Box; and I do hereby declare the following to be a full, clear, and exact description of the same, reference being had to the accompanying drawing forming part of this specification, in which–*

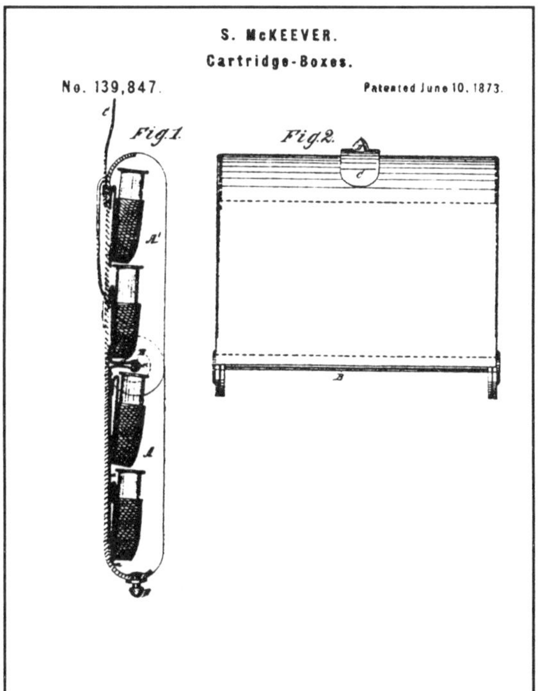

Photo 1.

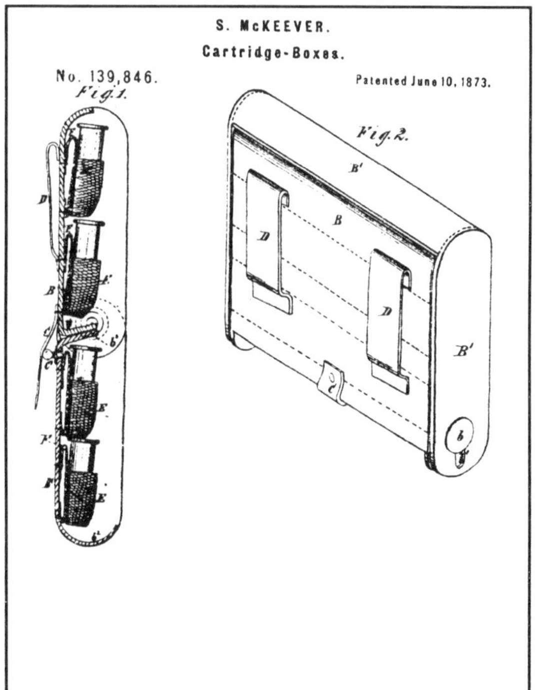

Photo 2.

Figure 1 is vertical section showing the box open. Fig. 2 is a front view.

The invention relates to the construction of that class of cartridge-boxes which are made in two parts hinged together.

The invention will first be fully described in connection with all that is necessary to a full understanding thereof, and then clearly pointed out in the claims.

In the drawing, A A' represents the two parts of a cartridge box, in which the cartridge-holders are attached thereto by leather or other flexible fabric. These two parts are connected by a pliable strip of leather, a, and provided with a pivot-rod, B, which is on the outside of this leather and about the middle thereof. These parts would turn on the leather hinge, but the latter would tend to bulge outward and work disadvantageously if the rod did not compel them to divide evenly over itself. On this rod the part A turns, and is made sufficiently higher than the one A' to fold over it at the upper end and form a cover thereto. On one part, A' is a flap, C;, which is slotted to fasten over a button, D, on the part A. By this improved construction I obtain a very strong and durable cartridge-box, and one which needs no slotting or lifting to enable it to swing open.

Having thus described my invention, what I claim as new, and desire to secure by Letters Patent, is–

1. A cartridge-box formed of two parts hinged together, one made higher than the other so as to fold over it at top and allow it to be latched in the manner described.

2. The flexible connection a combined with the rod or pivot B arranged across the middle thereof, as and for the purpose set forth.

3. The flap C and button D applied at the top of a two-part cartridge box, and for the purpose specified.

SAMUEL McKEEVER,

Capt. United States Army.

Witnesses: B.R. PIERCE,
JAMES GILLETTE.

Actually, McKeever patented two cartridge boxes on 10 June 1873. The other box, Pat. No. 139,846, was for a cartridge box very similar to the one with which we are now so familiar, but it was opened by raising the face-cover vertically, then folding it down in typical McKeever fashion. As shown in Photo 2, this box was secured by a tab at the bottom.

Both McKeever patents had four rows of cartridges, held in cloth loops affixed to an expansible bellows that allowed the cartridge bases to tip forward for easy grasping.

Army Trial and Adoption of the McKeever Cartridge Box

After the unsatisfactory trial of the 1872 Infantry Equipments, the Army scheduled another equipment board in 1874 to review proposals and equipment to improve the efficiency of infantry troops–particularly in field service. Additionally, the Board was to review all proposals and equipment in light of the newly-adopted .45/70 rifle cartridge.

The Board of five Regular infantry officers was composed of: LTC William R. Shafter, 24th Infantry; LTC A. McD. Cook,

Photo 3.

10th Infantry; LTC Thomas C. English, 2nd Infantry; MAJ Alexander Chambers, 4th Infantry and CPT M.H. Stacey, 12th Infantry. It met at Leavenworth Arsenal on June 1, 1874, and served until adjournment on Dec. 2, 1874.

A number of infantry-related equipments were reviewed, from knapsacks to cleaning kits, brace systems to bayonet scabbards. In particular, strong consideration was given to a British-style brace system to which a newly-approved cartridge box or boxes could be added. This brace system and appendages, submitted by First Lieutenant George H. Palmer of the 16th Infantry, was eventually recommended by the Board.

Early in the proceedings of the Board, First Lieutenant McKeever submitted his newly-patented cartridge box in a 40-round configuration.

The 40-round McKeever Cartridge Box was, like the Hagner No. 1 Cartridge Box of two years prior, too tall to allow a soldier to bend forward comfortably. For this and possibly other reasons, the Army Board asked that a smaller version be made for trial. Both McKeever and Watervliet Arsenal provided reduced-size boxes and, eventually, the pattern that collectors know as the First Pattern was adopted to go with the new brace system. Photo 3 shows the 1874 Ordnance Memorandum No. 19 depiction of the Palmer Infantry Brace System with its two integral First Pattern McKeever cartridge boxes in the new .45/70 calibre. The new, reduced-size cartridge box was compact, attractive, sturdy and space-efficient. And, it was no more costly to produce than competitive designs.

Despite the deliberations and experience of the 1874 Infantry Board, the new brace system was not well recieved by the field infantry. Like their brothers in the cavalry, the infantry soldier found the new, 1874 Infantry Brace System as hot and irksome as the cavalry had their trial 1872 Brace System. Both systems were highly unpopular and their use gave way to the concept of the "scouting", "hunting" or, as we know it today, the "prairie" belt as the much-preferred means of carrying ammunition in the field.

By 1876, even the Ordnance Dept. itself conceded that, for field service, the prairie belt was the method of choice for ammunition carriage in the field and began production of belts for combat branches. However, the compact and handsome McKeever Box continued to be favored for garrison use and it saw production and use from 1874 through 1910 in four rifle cartridge calibres, four models and production at two government arsenals–not to mention production by at least two commercial producers.

The McKeever Cartridge Box adopted, as indicated by Photo 3, opens at the top, and the face folds forward 180 degrees–exposing two interior rows of cloth loops, each row containing ten rounds. When the box was open, each cartridge row was presented to the user base up, each cartridge secured firmly by a wide canvas or web loop.

The First Pattern McKeever Box is called by collectors, the "Flat Side", due to its absence of a special tool pocket or other provision. This absence was noted by the Ordnance Office and all succeeding patterns of McKeevers had a tool pocket in a bulge on the left end of the box. It is felt this feature was incorporated in 1876, and coincided with the adoption of the M1876 Combination Tool. In form and function, this pocket followed the designs earlier used in the Hagner No. 1 and 2 Cartridge Boxes–to include a short, interior tool-securing strap.

A simple, sturdy axial brass rod along the bottom of the box not only serves as a box stiffener, but its ends serve as the two hinges around which the box cover pivots open. This feature is common to all McKeevers.

The McKeever Cartridge Boxes generally can be categorized in four basic patterns–though some other very interesting variations will be shown in the following section. The four patterns are as follows:

First Pattern–the original 1874 Trial box with flat sides (no tool pocket), leather bellows and a brass wear escutcheon on the closing tab. The early style of this box had no brass corner staples, but this feature was introduced in later First Pattern manufacture. Loops stitched at the bottom and riveted at the top.

Second Pattern–Adoption of a tool pocket in the left end of the box–all other features similar to the later First Pattern in the early boxes. Late boxes have canvas bellows, but retain brass tab escutcheon and corner staples.

Third Pattern–Deletion of the brass corner staples and tab escutcheon and adoption of canvas bellows.

Fourth Pattern–Always seen in russet and somewhat reduced in length, due to a reduction of the calibre (from .45/70 to .30/40 Krag.) Loops riveted on.

The change from leather to cotton bellows (Second to Third Pattern) was due to the formation of verdigris where the copper cartridges touched the leather. Since the area of contact for each cartrdige ran well down into the back of the loop (and, thereby, the cartridge case), the formation of verdigris posed a serious problem to chamber fouling and extraction until the bellows material was changed.

The .50 calibre McKeever Boxes were manufactured by the Army for the various state militias–all of whom were armed with Models 1866 Rifle Musket or M1868 or M1870 Rifles. This manufacture took place after the McKeever had been adopted and produced in .45/70. Interestingly, this production of .50 calibre McKeevers took place after the Army had modified its inventory of Hagner Boxes from .50 calibre to .45 calibre!

McKeever Cartridge Boxes were not manufactured for shotshells, but at least two experimental pistol calibre boxes were manufactured but not adopted.

The shape of the tool pocket bulge on the box's left end varied over its period of production and with the producer. The actual construction of the front or face portion of the box varied also. Later boxes were of pieced construction, i.e. the face was one piece, stitched to two end caps. The other variation, earlier and much less common, is of folded construction, in which the face folds around toward the back, forming the two end caps.

The measurements used are merely for identification. OAT (overall thickness) does not include the two belt loops on the back of the box body.

First Pattern McKeever Cartridge Box (Early and Late Styles)

Calibre: .45/70 Government

Capacity: 20 rounds in two rows of ten rounds each.

Measurements:

OAL–	6 1/2"
OAW–	3 7/8"
OAT–	1 5/8"

Colour/Materials/Construction: A black leather cartridge box and belt loops. The face section of this box is of folded construction, and the closing tab has a brass escutcheon. The face section of the earliest style of this box does not have brass corner staples. The later style of the First Pattern McKeever introduced two brass upper corner staples.

Varnished canvas loops are stitched to fair leather bellows.

Markings: The lower bellows are stamped with two sets of inspectors' initials: **"A.R.S."** and, underneath, **"C.M."**

"WATERVLIET ARSENAL" is stamped in two lines on the base of the closing tab.

Comments: These First Pattern McKeevers were only made at Watervliet Arsenal and were issued in pairs with the 1874 Palmer Infantry Brace System. These early McKeevers are very desirable, and the Early Style (without corner staples) is rare.

Photo 1.

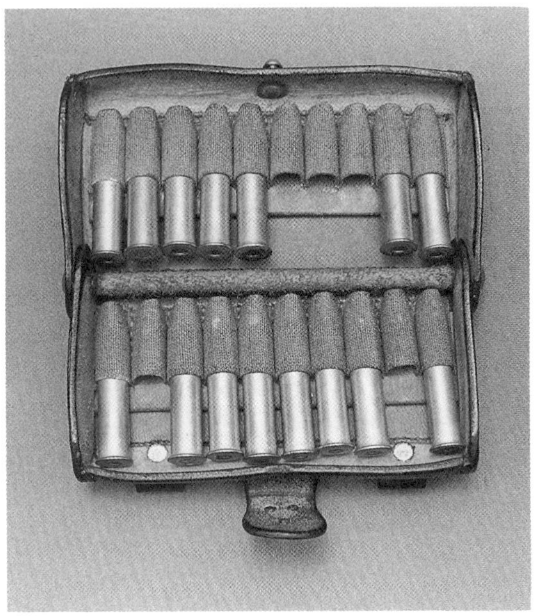

Photo 2.

The "Cut Loop" McKeever Variation

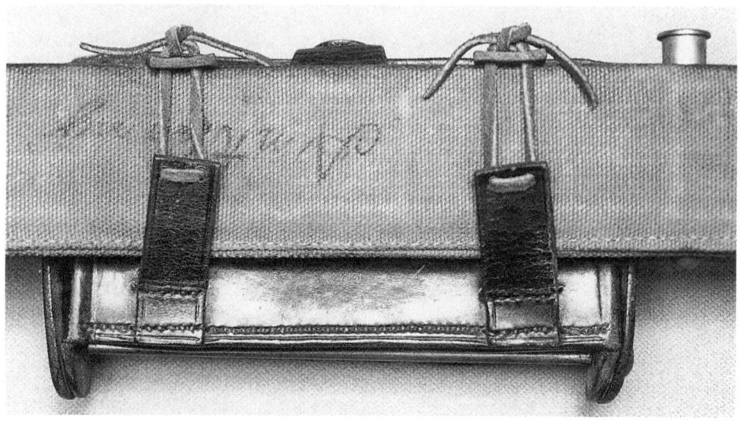

Photo 1.

The McKeever Box was first issued in 1874 as part of the unsuccessful Palmer (Infantry) Brace System. After the demise of the brace system, the popular, compact McKeever continued in service. In 1876, the Army adopted the leather and canvas cartridge belt or "prairie" belt–the first of many variations for infantry and cavalry. The earliest variation of this belt was narrower than the later prairie belts issued, and its narrow width allowed the lead bullets of the long .45/70 cartridges to rub and mark the soldiers' uniforms. Happily, however, the narrow belt was just narrow enough to accept the mounting of the standard McKeever Box (First Model) when the loops were cut, punched for holes and attached with leather thongs–as shown.

When mounted on a cartridge belt, the soldier would give up two or three rounds covered by the box loops, but he would gain the advantage of an additional 20 rounds of ammunition in a very compact space.

From the specimens of cut-loop McKeevers examined, it appears that they were done with great uniformity–suggesting that they were done at one well-equipped and central location, perhaps even an arsenal.

Specimens of these early Indian War modified McKeevers are very desirable and are quite scarce. Unfortunately, none have yet been found with authentic unit marks.

Photo 1 shows the typical mounting (reverse) of a Cut Loop First Model McKeever on a First Pattern 1876 Prairie Belt. Photo 2 shows the McKeever from the front. Note that the belt was painted blue (the infantry uniform colour) on the exterior–a feature occasionally seen, but not common.

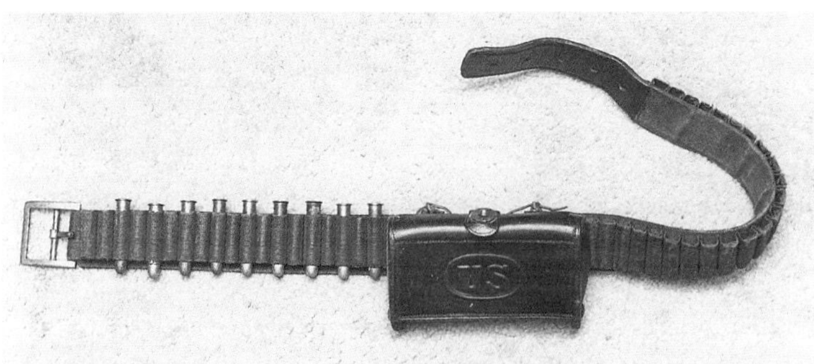

Photo 2.

Addition of the Tool Pocket

Photo 1.

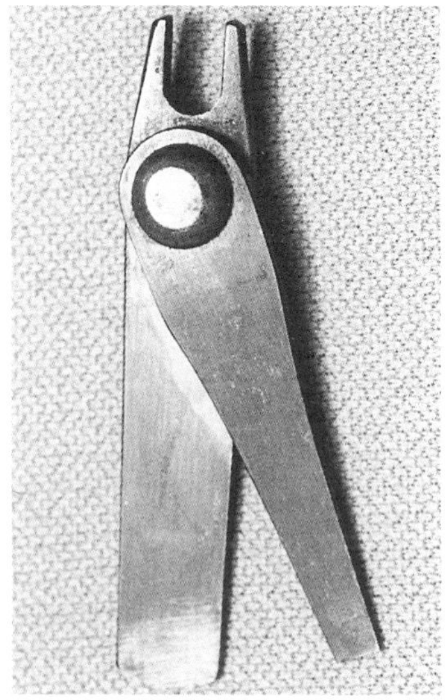

Photo 2.

The original intention of the Army Ordnance and the 1874 Infantry Board was that a rifle maintenance tool would be carried in a cartridge loop, if necessary. Probably at the field level, this loss of one cartridge (in a box of twenty) was not felt to be acceptable, and a design change to the left box end was adopted. The change required a wet-forming of a bulge to both left box ends and the addition of a small, leather cross strap for securing the tool in the pocket (bulge). This new design, incorporating a tool pocket constitutes a Second Pattern McKeever. Photo 1 shows a First Pattern McKeever on the left and a Second Pattern (with tool pocket bulge) on the right.

The tool for which this pocket was designed was the Model 1876 Combination Tool, shown in Photo 2. This compact little tool would be carried in all succeeding patterns of McKeevers, and offered a mainspring clamp and two screwdrivers. It was modified for Indian War service from either the M1870 Combination Tool or the M1863 Combination Tool. Photo 3 is an inside view of the tool pocket adopted in the Second Pattern McKeever with the short cross strap retaining the 1876 Combination Tool.

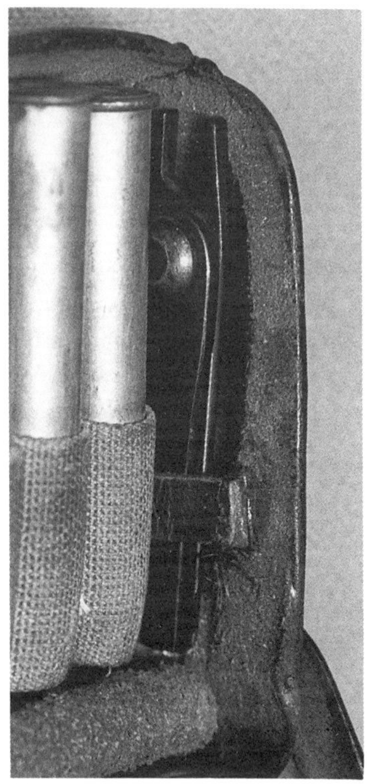

Photo 3.

McKeever Cartridge Box Tool Pocket Configurations

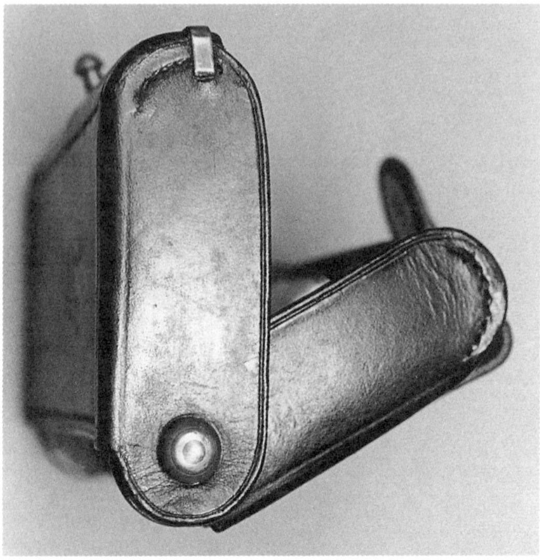

Photo 1.

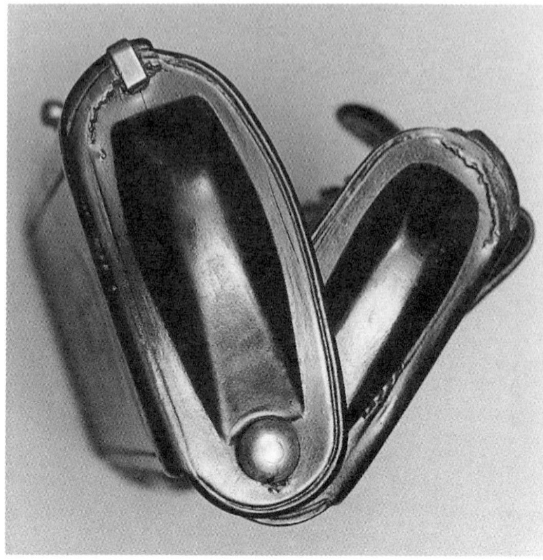

Photo 2.

The bulge wet-pressed into the right end of the 2nd through 4th Pattern McKeever Cartridge Boxes, differed over time of production and producer. The four most distinct forms are shown here.

Photo 1 shows a Watervliet Arsenal-produced First Pattern McKeever. Called the "Flat Side" by collectors, it is the original production piece, and had a folded front and no provision for a tool pouch. The tool was expected to be slipped into an end cartridge loop and was easily retained in this manner, but could be jolted loose.

Photo 2 shows the smooth, rounded bevel of a Second Pattern McKeever, produced by Rock Island Arsenal.

Photo 3 shows the more flattened, rounded and beveled bulge form. This is found on a commercially-manufactured "NGP" McKeever.

Photo 4 shows the late, elongated and rounded tool pocket bulge found on a Third Pattern McKeever, manufactured by Rock Island Arsenal. This box has the late, block style RIA stamp and a brass, oval "NJ" plate. This style of bulge continued through the production of the russet, 4th Pattern McKeever.

Photo 3.

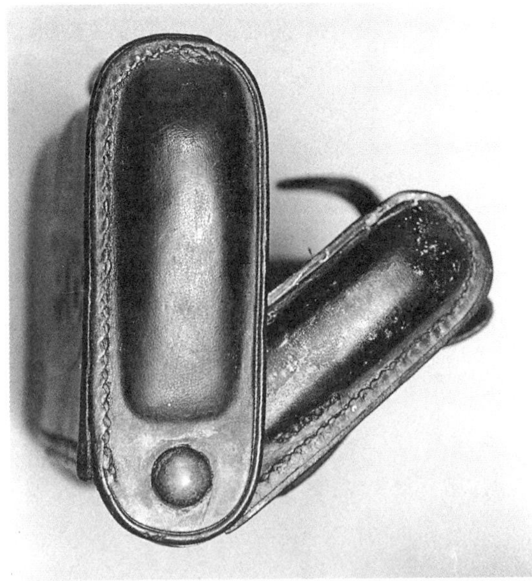

Photo 4.

Second Pattern R.I.A. McKeever Cartridge Box

Calibre: .45/70 Government

Capacity: 20 rounds in two rows of 10 rounds each

Measurements:

OAL–	8 13/16"
OAW–	4"

Colour/Materials/Construction: A black, folded front box with the standard, Second Pattern tab escutcheon and top corner staples. Fair leather bellows and varnished, dark canvas cartridge loops.

The tool pocket bulge is distinctly beveled. The belt loops are stitched at the bottom and riveted at the top.

Markings: The base of the closing tab has the 1874-period **"ROCK ISLAND ARSENAL"** stamp that has been called the "sunrise stamp" because of its rising sun appearance.

The lower leather bellows is stamped with the inspector's initials, **"C.McM."**

Photo 1.

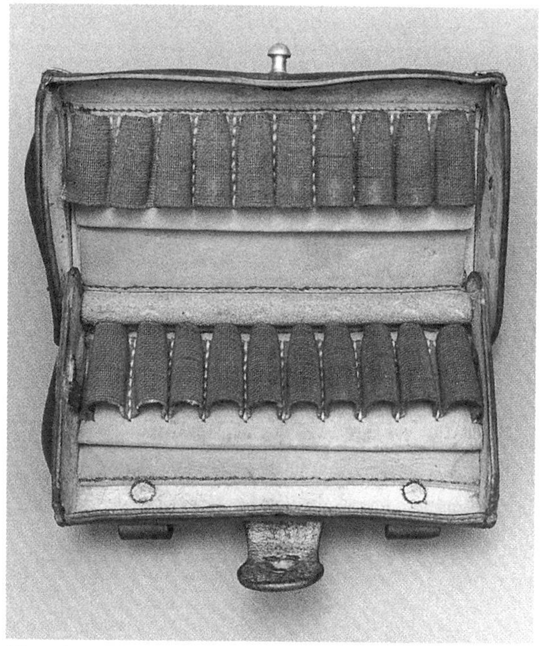

Photo 2.

Second Pattern R.I.A. McKeever Cartridge Box

Calibre: .45/70 Government

Capacity: 20 rounds in two rows of 10 each.

Measurements:

OAL– 7"
OAW– 4"

Colour/Materials/Construction: A black, folded front box with standard Second Pattern tab escutcheon and top corner staples. Almost identical to the preceding Second Pattern RIA McKeever, this box has very light-coloured hard canvas cartridge loops that also measure about 1/16" narrower than the other box.

Both boxes have a beveled tool pocket bulge.

Markings: The base of the closing tab has the 1874-period **"ROCK ISLAND ARSENAL"** sunrise stamp.

The inspector's stamp "W.D.B." is stamped on the left of the lower bellows.

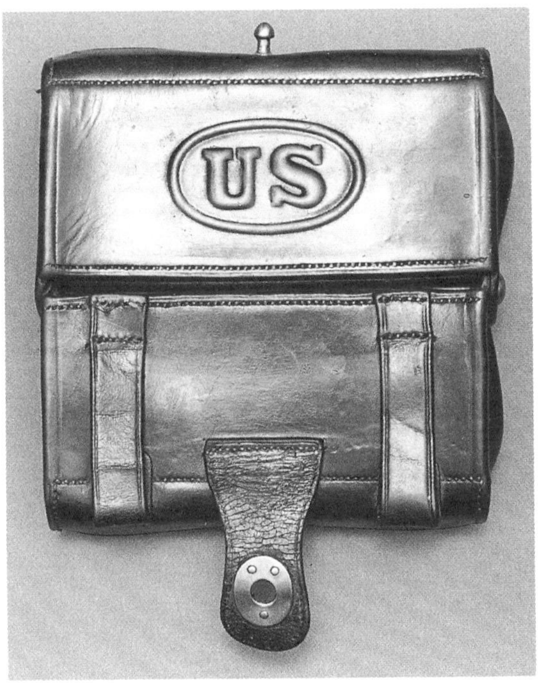

Photo 1.

Photo 2.

Second Pattern Watervliet Arsenal McKeever Cartridge Box

Photo 1.

Photo 2.

Calibre: .45/70 Government

Capacity: 20 rounds in two rows of ten rounds each

Measurements:

OAL–	6 7/8"
OAW–	3 29/32"

Colour/Materials/Construction: This black leather box is of the standard stitched construction with fair leather bellows, varnished canvas loops sewn to the bellows, brass corner staples and tab escutcheon.

Markings: The lower end of the closing tab (just above the stitching) is stamped **"WATERVLIET ARSENAL"** in two lines. The lower bellows is marked by two inspectors: **"A.R. SMITH"** and **"H.C."**

Second Pattern R.I.A. McKeever Cartridge Box

Calibre: .50/70 Government

Capacity: 20 rounds in two rows of 10 rounds each.

Measurements:

OAL–	7 15/16"
OAW–	3 7/8"
OAT–	1 3/4"

Colour/Materials/Construction: A standard, stitched construction, black leather box body and loops. Brass tab escutcheon and corner staples.

Varnished canvas loops 1 1/4" in height, mounted on fair leather bellows.

Markings: The 1874 period R.I.A. mark is stamped on the base of the closing tab. Inspector's initials **"D.S.L."** and **"J.O.G."** are stamped on the lower right side of the lower bellows.

Comments: This is a very scarce and desirable McKeever. This may well be the first pattern McKeever in .50 calibre and, though marked **"U.S."** on the cover, may have been the first militia McKeever. .50 cal. McKeevers were not made for Regular Army issue.

Photo 1.

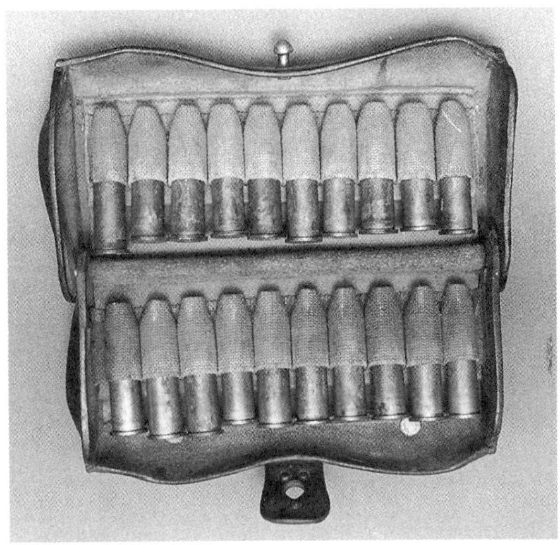

Photo 2.

Early Third Pattern Unmarked McKeever Cartridge Box

Calibre: .45/70 Government

Capacity: 20 rounds in two rows of ten rounds each.

Measurements:

OAL–	7"
OAW–	4 1/4"

Colour/Materials/Construction: A standard, stitched construction, black leather box with two corner staples but manufactured without a brass tab escutcheon. The box has dark canvas bellows and cartridge loops.

Markings: Unquestionably arsenal-manufactured, but unmarked. No inspector's initials and only the standard "US" in an oval on the cover.

Comments: Quite unusual in its absence of arsenal and inspector marks. Corner staples indicate early manufacture.

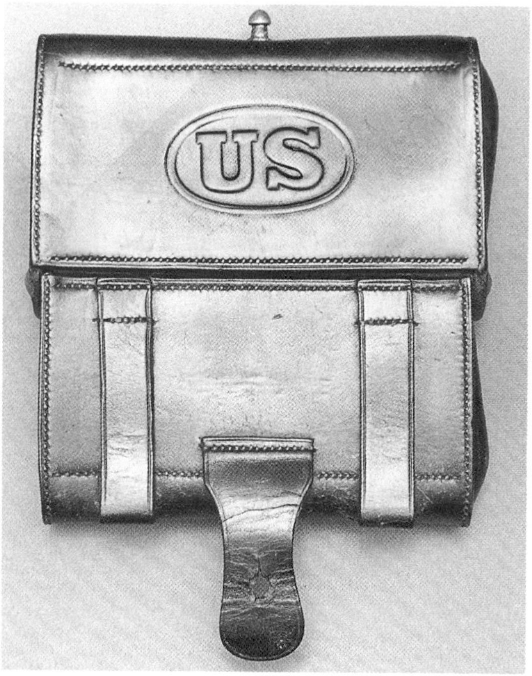

Photo 1.

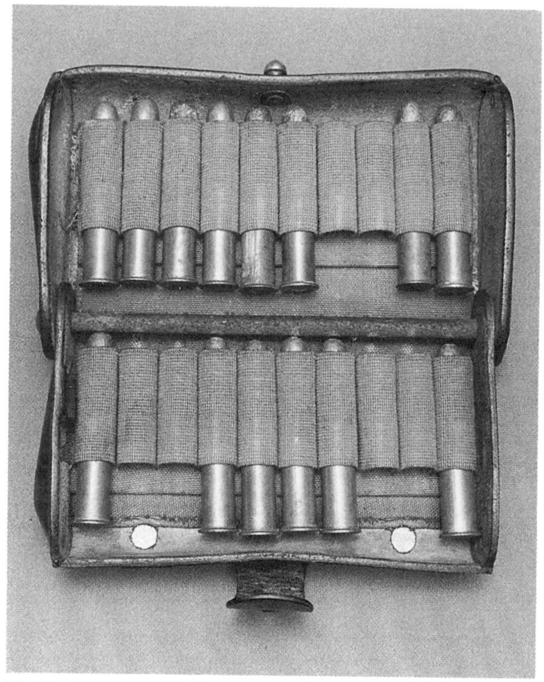

Photo 2.

1881 Mills Second Pattern McKeever Cartridge Box

Calibre: .45/70 Government

Capacity: 20 rounds in two rows of 10 cartridges each

Measurements:

OAL–	6 7/8"
OAW–	4 1/4"

Colour/Measurements/Construction: A standard, stitched construction, black leather box body and loops. Brass tab escutcheon and corner staples. What sets this box aside from other McKeevers is its use of Mills Woven Cartridge Belt Co. web bellows and integral (woven) loops. These components were procured by the government separately and incorporated by Watervliet Arsenal into this cartridge box.

Centered on each bellows is a small brass tag 1" x 1/4", embossed with a decorative edge and the following, in two lines:

"ANSON MILLS
PAT. DEC. 28 80"

The 1 9/16" web loops are integral with the web bellows. The entire height of the bellows (on left and right sides) are edged with a light coloured, narrow basting of leather (to preclude fraying).

Markings: **"WATERVLIET ARSENAL"** is stamped in two lines at the base of the closing tab.

Sharply stamped on the inner side of the right hand, inner box is the inspector's stamp **"A.R. SMITH".**

Comments: The use of the Mills webbing was essentially an Army trial and production quantities were low. The .45/70 Mills McKeever was a regulation, Regular Army issue box and is very desirable today.

Photo 1.

Photo 2.

1881 Mills Second Pattern McKeever Cartridge Box

Calibre: .50/70 Government

Capacity: 20 rounds in two rows of 10 rounds each

Measurements:

OAL–	7 3/4"
OAW–	3 5/8"

Comments: The construction and markings of this box are similar to the preceding .45/70 Mills McKeever. What is significant is that it was manufactured in .50 cal. for state militia use, since most states in 1881 still had the older .50 calibre rifles M1866, M1868 and M1870.

Though unclear in Photo 1, the cover embossing is the standard **"U.S."** in an oval. The overall dimensions of this box are different from the .45/70 box, indicating different manufacturing templates and, probably, different production dates. Though undoubtedly of arsenal manufacture, it was never arsenal nor inspector-marked.

This is a very desirable McKeever variation.

Photo 1.

Photo 2.

Cavalry-Used McKeevers

Both the cavalry and infantry preferred the "scouting" or prairie belt for field use, but both branches favored the trim, brass mounted McKeever for garrison duty–accounting for the McKeever's production well past the Indian War. The 1878-1879 Special Board of officers convened by the Army Adjutant General specifically recommended that the Dyer Carbine Cartridge Pouch, the pouch for pistol cartridges and the Hazen (Carbine Cartridge) Loops "be replaced by the McKeever cartridge-box for use in garrison and by the double cartridge belt for field service." This, essentially, became the pattern of use for the Army for the next two decades.

The box in Photo 1 is double-stamped **"CO.L. 5 CAVY"** and the box in Photo 2 is stamped **"D 9 CAV"**. Note that both boxes are the Second Pattern McKeever.

Photo 1.

Photo 2.

Third Pattern Watervliet Arsenal McKeever Cartridge Box

Calibre: .45/70

Capacity: 20 rounds in two rows of 10 rounds each.

Measurements:

OAL–	6 13/16"
OAW–	4 1/16"
OAT–	1 11/16"

Colour/Materials/Construction: A standard, stitched construction, black leather box body. Varnished canvas loops 1 17/32" high on canvas bellows.

Markings: "WATERVLIET ARSENAL" is stamped in two lines on the base of the closing tab. No inspector's marks show, even though the condition of the leather is excellent.

Photo 1.

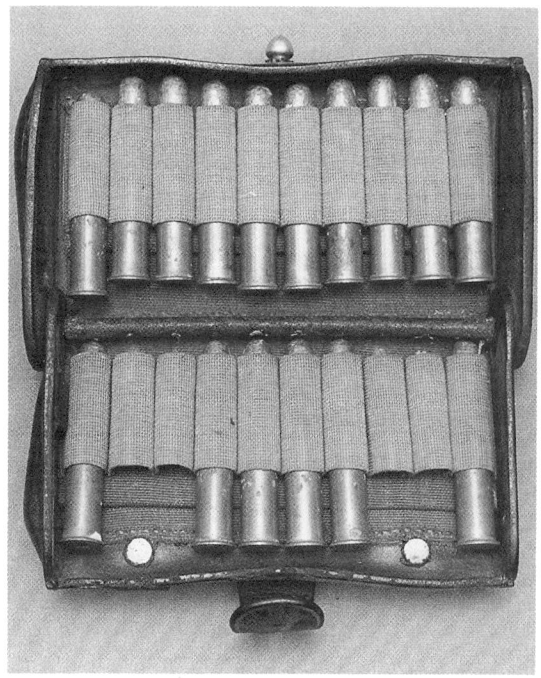

Photo 2.

Third Pattern Watervliet Arsenal McKeever Cartridge Box

Calibre: .45/70 Government

Capacity: 20 rounds in two rows of ten rounds each.

Measurements:

OAL–	6 7/8"
OAW–	4"

Colour/Materials/Construction: A standard stitched construction box with soft canvas bellows and cartridge loops.

Markings: A standard **"U.S."** is embossed on the corner of the box. **"WATERVLIET ARSENAL"** is stamped in two lines on the base of the closing tab. The inspector's initials **"C.H."** are stamped on the inner right end of the box. A unit stamp (box number) **"60"** is stamped on the closing tab.

Photo 1.

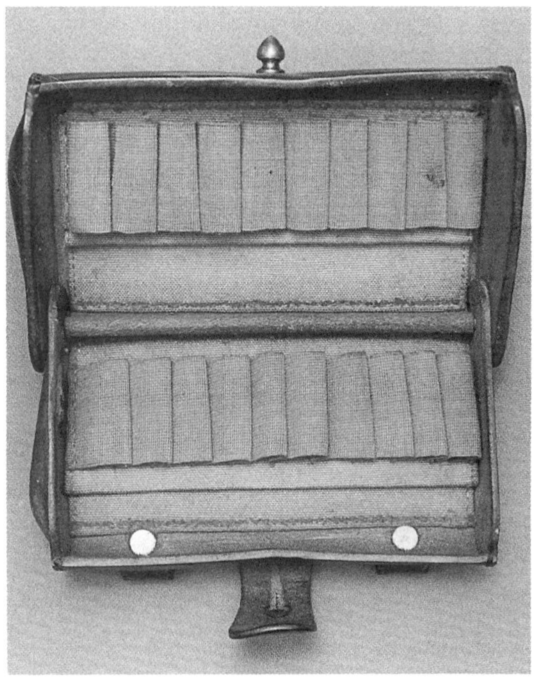

Photo 2.

Third Pattern Watervliet Arsenal Cadet McKeever Cartridge Box

Calibre: .45/70 Government

Capacity: 20 rounds in two rows of 10 rounds each

Measurements:

OAL–	8 7/8"
OAW–	3 7/8"
Buckles–Height	1 1/8"
Width	2 3/8"

Colour/Materials/Construction: Excepting the brass buckles on the back, this is a conventional black, stitched construction box. It has soft canvas cartridge loops and bellows.

The twin, slotted brass buckles are secured at their base by stitched leather chapes. Above them, and unique to the Cadet McKeever, are two horizontal leather loops, which serve to secure the cloth shoulder strap and keep the box hanging flat against the body. Photo 3 shows the proper shoulder strap threaded through the buckles and up through the leather loops. The tool pouch bulge is rounded.

Markings: "WATERVLIET ARSENAL" is stamped in two lines on the base of the closing tab.

Stamped at the very top of the inner section of the box is, **"Co. A No. 20"**, a cadet unit mark that could have been applied at West Point.

Photo 1.

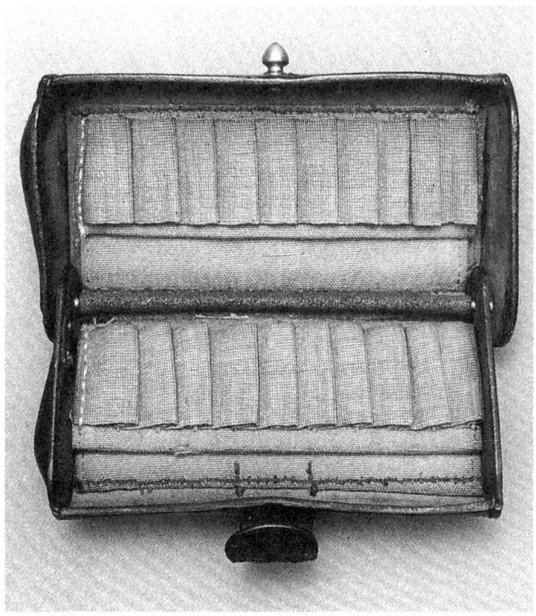

Photo 2.

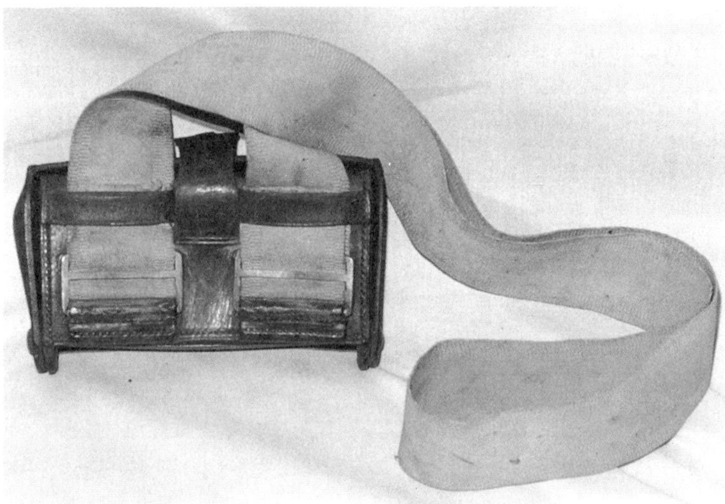

Photo 3.

Fourth Pattern R.I.A. McKeever Cartridge Box

Calibre: .30/40 Krag-Jorgenson (.30 Government)

Capacity: 20 rounds in two rows of 10 rounds each

Measurements:

OAL–	6 1/4"
OAW–	4 1/2"

Colour/Materials/Construction: A standard, stitched construction leather box and loops, but one in "fair" or russet leather. Reduced in length due to the size of the new cartridge, the box (when closed) appears almost square.

The cartridge loops and bellows are a tan canvas. Note that the loops (11/16" in width) are at the bottom of the box and only secure the bullet and part of the cartridge case neck.

The belt loops are riveted on with flat head rivets top and bottom. The bottom of the loops are also secured with a line of stitching. The tool pouch bulge is rounded.

Markings: The exterior of the right inner box end is stamped, **"ROCK ISLAND ARSENAL"** over **"1909"** over **"H.E.K."** (an inspector). **"18"** is stamped on the tip of the tab.

Comments: All of these Fourth Pattern McKeevers are made for .30/40 Krag ammunition and were probably meant for the state militias, armed with Krag Rifles and Carbines. It is noteworthy that the Regular Army Cavalry was armed with Krag Carbines until 1911.

This box was not intended for combat use at this late date, but was intended for garrison guard, parades, military schools, etc. The fair or russet colour commenced in 1903 for all U.S. accoutrements.

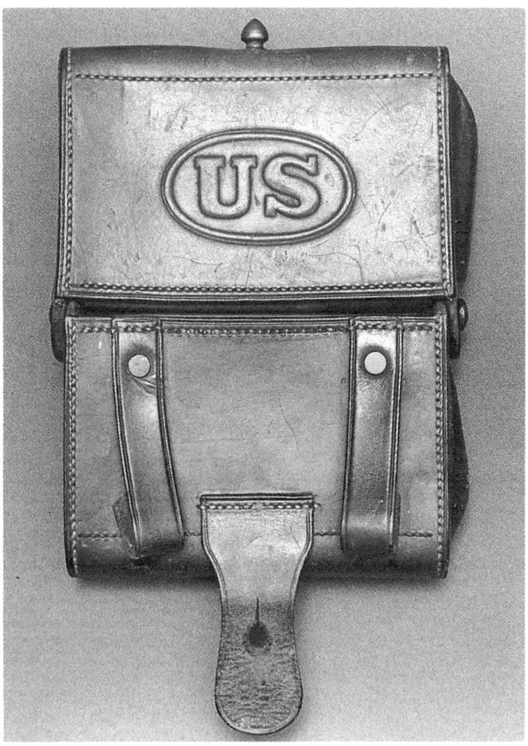

Photo 1.

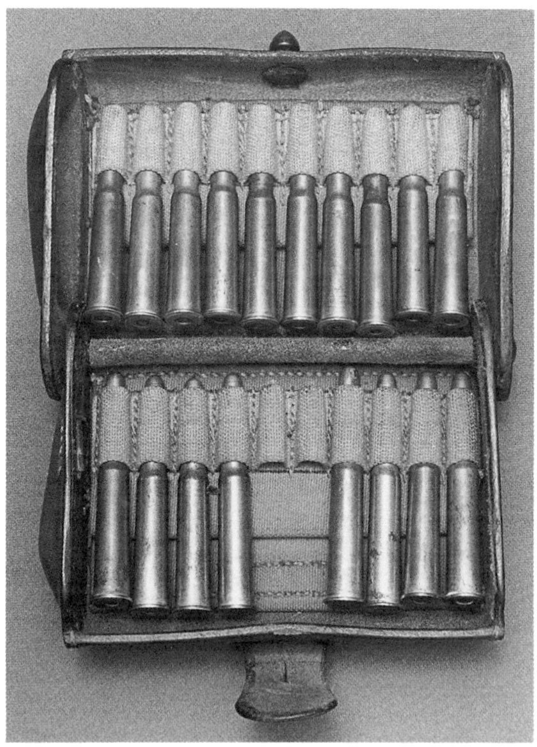

Photo 2.

Experimental 12-Round Pistol McKeever Cartridge Box

Calibre: .45 Army Revolver (.45 Schofield)

Capacity: 12 rounds in two rows of six cartridges each.

Measurements:

OAH– Body without Loops nor Finial	3 1/2"
OAW–	4 1/4"
OAT–	1 1/2"
Height of Cartridge Loops	1 1/16"

Colour/Materials/Construction: A reduced size Third Pattern McKeever of stiff black leather. The bellows are a light tan canvas and the cartridge loops themselves are white canvas.

Markings: The tab is stamped on its base, **"ROCK ISLAND ARSENAL."** The lower base of the box bears two sets of inspectors' marks: **"WTG"** over **"F.A.B."**

Comments: The entry in the 1909 Rock Island Arsenal catalogue for this item reads as follows, **"Cartridge Box. McKeever pattern for Cal. .45 revolver ammunition, for artillery sabre belt. Made at Rock Island Arsenal following instructions from Chief of Ordnance, September 27, 1894."** Two years later, the Army would adopt a simpler, less costly cartridge box for the new .38 calibre Army revolver.

Photo 1.

Photo 2.

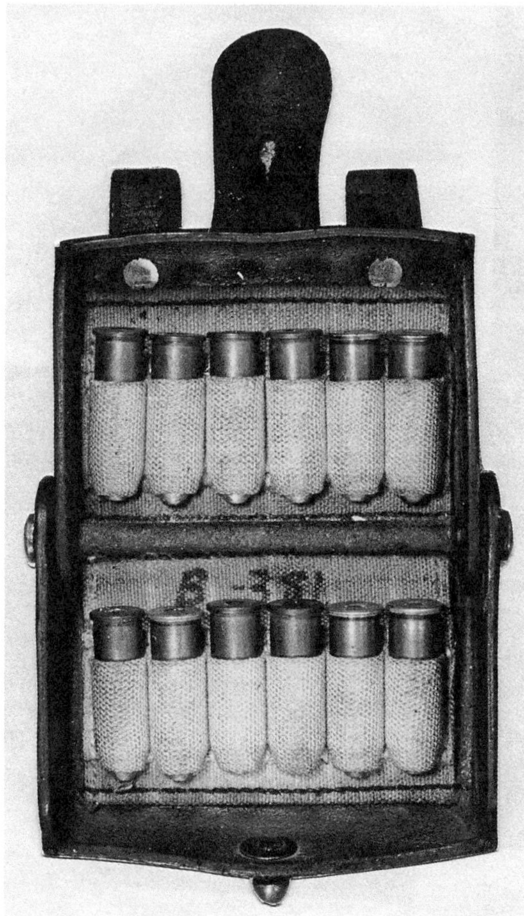

Photo 3.

Experimental 20-Round Pistol McKeever Cartridge Box

Calibre: .45 Army Revolver (.45 Schofield)

Capacity: 20 rounds in two rows of 10 cartridges each.

Measurements:

OAH– without Loops nor Finial 2 7/8"
OAW– 6 1/2"
OAT– 1 1/2"

Colour/Materials/Construction: A variant Third Pattern McKeever of stiff black leather. The bellows and cartridge loops are a light tan canvas.

Markings: With the exception of the U.S. embossed in an oval on the cover, the box is unmarked.

Comments: The entry in the 1909 Rock Island Arsenal catalogue for this item reads as follows, **"Cartridge Box. McKeever pattern, proposed to carry 20, Cal. .45 revolver cartridges. Made at Rock Island Arsenal, July 20th, 1989."**

What is interesting is that the number of cartridges proposed by this concept reflects a definite combat roll for the .45 cal. revolvers in the hands of the artillery and any other troops so armed. Interestingly, too, is the continued Ordnance interest in a cartridge box dedicated to this obsolescent round two years after the Army adoption of the M1896 Pistol Cartridge Box. (The M1896 was used with a bored wooden block for .38 cal. cartridges and with the 12 round cardboard box holding .45 rounds.)

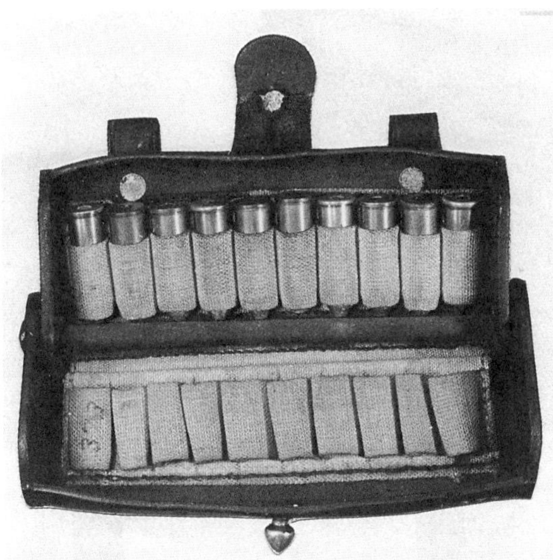

Photo 1.

Photo 2.

Photo 3.

"K.S.G." (Kentucky State Guard) Second Pattern McKeever Cartridge Box

Calibre: .45/70 Government

Capacity: 20 rounds in two rows of 10 rounds each

Measurements:

OAL–	6 1/2"
OAW–	3 1/2"
OAT–	1 1/2"

Colour/Materials/Construction: A standard, stitched construction, black leather box with dark canvas loops stitched onto fair leather bellows. The box is fitted with standard Second Pattern brass tab escutcheon and upper corner staples.

However, on the back of the box and secured to the belt loops are two period, black iron buttons. These are typical state militia modifications which would allow the box to be mounted on a shoulder sling and carried on the side, under one arm. Such a use would have only parade/ceremonial justification. The sling could be cloth, leather or a combination of both materials.

Markings: This box is stamped with two sets of inspectors' marks inside, on the fair leather bellows: **"D.S."** and **"A.R. SMITH"**. In addition, the base of the closing tab is marked **"WATERVLIET ARSENAL"**.

Photo 1.

Photo 2.

Photo 3.

"K.S.G." (Kentucky State Guard) Third Pattern McKeever Cartridge Box

Calibre: .45/70 Government

Capacity: 20 rounds in two rows of 10 rounds each

Measurements:

OAL–	6 3/4"
OAW–	4 3/8"

Comments: An otherwise standard, Watervliet Arsenal-manufactured Third Pattern McKeever. This box was made for the Kentucky militia.

"N.C." (North Carolina) Third Pattern McKeever Cartridge Box

Calibre: .50/70 Government

Capacity: 20 rounds in two rows of 10 rounds each

Measurements:

OAL–	6 7/8"
OAW–	4 1/4"

Comments: An otherwise standard, Watervliet Arsenal-manufactured Third Pattern McKeever. This box was made for North Carolina militia.

"NG" (National Guard) Third Pattern McKeever Cartridge Box

Calibre: .50/70 Government

Capacity: 20 rounds in two rows of 10 rounds each.

Measurements:

OAL–	7 5/8"
OAW–	4"
OAT–	1 11/16"

Colour/Materials/Construction: A standard stitched construction, black leather box with dark, varnished canvas loops on canvas bellows. The stitching between the loops is a very light ivory colour.

Markings: "WATERVLIET ARSENAL" is stamped in two lines at the base of the closing tab. The inspector's initials **"J.B."** are stamped on the inside end cap on the right side of the box.

Comments: A very desirable McKeever, this box demonstrates the length of time state militias continued to use the .50 calibre rifle muskets and rifles–long after the Army adoption of the .45/70 series "Trapdoor" Rifles. The .50 calibre Third Pattern McKeever Box is a scarce cartridge box.

Photo 1.

Photo 2.

"N.G." (National Guard) McKeever Cartridge Box–Third Pattern

Calibre: .45/70 Government

Capacity: 20 rounds in two rows of 10 rounds each

Measurements:

OAL–	6 7/8"
OAW–	4 1/4"
"N.G."– Height	1 15/32"
Width	1 5/8"

Colour/Materials/Construction: A standard, stitched construction black leather box body and loops. Soft, tan canvas loops and bellows.

The current **"N.G."** brass plate is secured through the box face with four fine brass wires, folded over. This emblem may not be original to the box (certainly something has been removed prior), but the box has always been a militia box due to its lack of a "U.S." embossing. Possibly, the original insignia snagged and tore loose–then was replaced.

The belt loops are stitched at the bottom and riveted at the top. The tool pocket bulge is rounded.

Markings: "WATERVLIET ARSENAL" is stamped in two lines on the base of the closing tab. The exterior of the right, inner box end is stamped, **"C.H."** (an inspector).

Photo 1.

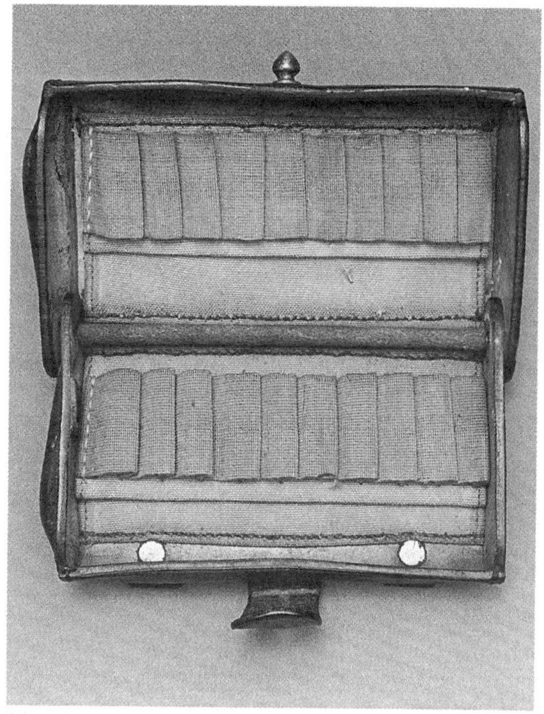

Photo 2.

"N.G.P." (National Guard of Pennsylvania) McKeever Cartridge Box

Calibre: .45/70 Government

Capacity: 20 rounds in two rows of 10 rounds each.

Measurements:

OAL–	6 5/8"
OAW–	3 3/4"

Colour/Materials/Construction: A commercially-made, standard stitched construction black leather box body and loops, but unusual in several respects. First, the face is deeply struck with the "N.G.P." mark. Second, though the box lacks a tab escutcheon and corner staples, it has the early style, fair leather bellows. Third, the varnished canvas loops are specially woven in bright colours. The colour pattern of the loops is as follows: dark blue top and bottom bar; broad tan bars; narrower blue bars; very narrow yellow bars; and a red center stripe.

The leather of the box, particularly in the loops and tab, is lighter than in arsenal-produced McKeevers. The cutting of the leather and stitching is also not quite up to arsenal standards.

The ends of the brass hinge rod are rougher and have distinct, slightly raised centers. The belt loops are sewn at the bottom and riveted at the top. It has a distinctly beveled tool pocket bulge but no inner tool-retaining strap.

Markings: None

Photo 1.

Photo 2.

"N.G.P." (National Guard of Pennsylvania) Second Pattern Variant McKeever Cartridge Box

Calibre: .45/70 Government

Capacity: 20 rounds in two rows of 10 rounds each.

Measurements:

OAL–	6 11/16"
OAW–	3 7/8"

Colour/Materials/Construction:A black leather cartridge box and belt loops. The face section of this box is of folded construction. The closing tab has a brass escutcheon and the face section of the box has two brass corner staples.

The soft, white canvas loops have vertical lines of yellow stitching securing them to the fair leather bellows.

The markings of this box and the use of leather bellows argues for this being a Second Pattern box, but the lack of a brass tab escutcheon and corner staples places it in a "variant" category. These latter features were probably deleted by the civilian manufacturer and/or state procurement, as a cost-saving measure.

The belt loops and tab are of lighter leather than those of arsenal construction.

The tool bulge on the left end of the box, like the other **"NGP"** McKeever shown, has no tool-retaining strap on the inside.

Markings: Most noticeably, an embossed **"N.G.P."** in an oval is on the box face. The right end of the box is stamped **"D"** over **"47"** (D Company, 47th Pennsylvania).

The tab tip is stamped as follows (Photo 3):

Photo 1.

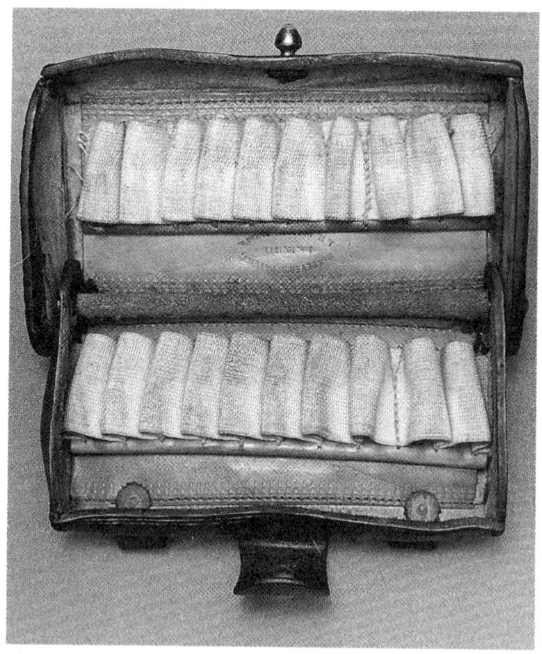

Photo 2.

McKEEVER'S PATENT
JAN. 10. 1873
J.H. WILSON PHILA

This mark is repeated on the lower bellows on the right side.

Photo 4 compares the cover stamps of the two N.G.P. boxes. The stamp at the top is from what is felt to be the earlier box–the one with the McKeever patent information. The lower box is the one with the striped loops.

Comments: It can be argued that this is a Third Pattern variant and the author cannot disagree, but the patent markings make me think this is an earlier box. An uncommon and desirable specimen.

Photo 3.

Photo 4.

"N.J." (New Jersey) Militia Third Pattern McKeever

Calibre: .45/70 Government

Capacity: 20 rounds in two rows of ten rounds each

Measurements:

OAL Box–	6 3/4"
OAW Box–	4 1/8"
OAL Plate–	2 11/16"
OAW Plate–	1 9/16"

Colour/Materials/Construction: A standard, stitched construction, black leather box with soft, light coloured canvas loops and bellows. The bellows are secured to the box body with dark stitching and the loops are stitched to the bellows with very light coloured ivory stitching.

A lead-filled, brass oval plate with the raised letters **"NJ"** was attached to the box face prior to attachment of the bellows and, so is original to manufacture of the box.

Markings: The later-style, large letter stamp of "ROCK ISLAND ARSENAL" is stamped in; two lines at the base of the closing tab. No inspector's marks are present on this box.

Comments: This is a very desirable box by virtue of its being one of very few manufactured as a plated militia box, i.e. not modified subsequent to manufacture. Another interesting feature is that the earlier style of decoration (an affixed brass plate) is used, rather than a leather embossing on this since it is a later production, Third Pattern box.

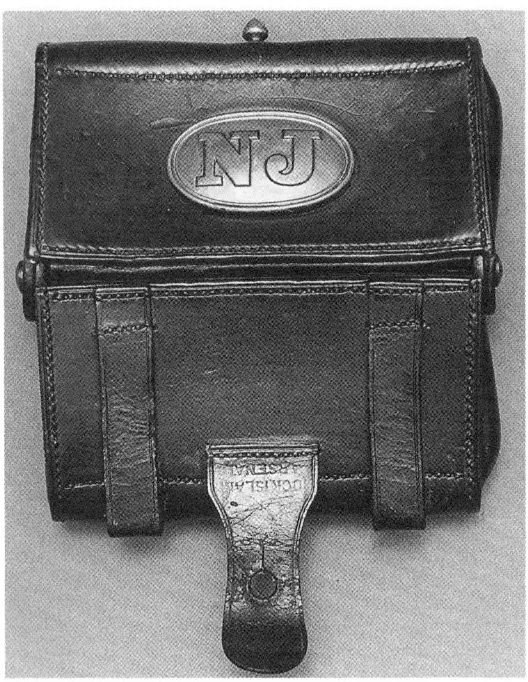

Photo 1.

Photo 2.

"U.S.M." Second Pattern McKeever Cartridge Box

Calibre: .45/70 Government

Capacity: 20 rounds in two rows of ten rounds each

Measurements:

OAL–	6 7/8"
OAW–	4 3/16"

Colour/Materials/Construction: A standard, stitched construction, black leather box with soft, dark canvas bellows and cartridge loops. The stitching between the loops is a very light ivory colour. Standard corner staples and closing tab escutcheon.

Markings: "WATERVLIET ARSENAL" is stamped in two lines at the base of the closing tab. The **"U.S.M."** embossed on the corner stands for U.S. Militia.

Comments: An unusual, early militia McKeever. It is probably a federal government attempt to provide a unique box to state militias that would not require the cost of a state-specific stamping die.

Photo 1.

Photo 2.

"U.S.M." Third Pattern McKeever Cartridge Box

Calibre: .45/70 Government

Capacity: 20 rounds in two rows of ten rounds each

Measurements:

OAL–	6 7/8"
OAW–	4 3/16"

Colour/Materials/Construction: A standard, stitched construction, black leather box with soft, dark canvas bellows and cartridge loops.

Markings: "U.S.M." embossed on the box cover in a standard oval. This stands for U.S. Militia, and was not specific to any one state militia. Inspector's initials "C.H." stamped on the left inner box end. "ROCK ISLAND ARSENAL", stamped in two lines at the base of the closing tab.

Comments: The U.S.M. McKeevers are quite desirable and appear to be a federal government attempt to provide a unique box to state militias that would not require the cost of a state-specific stamping die.

Photo 1.

Photo 2.

U.S.M.C. Third Pattern McKeever Cartridge Box

Calibre: .45/70 Government

Capacity: 20 rounds in two rows of 10 rounds each

Measurements:

OAL–	6 3/4"
OAW–	4"

Colour/Materials/Construction: A standard, stitched construction, black leather box body and loops. Soft canvas cartridge loops and canvas bellows. The belt loops are stitched on the bottom and riveted at the top.

Markings: Besides the **"USMC"** in a 3 9/16" x 1 9/16" oval on the cover, the base of the closing tab is stamped, **"ROCK ISLAND ARSENAL"** in two lines. No inspector's initials were apparent.

Comments: This scarce cartridge box probably predates the Third Pattern 6mm Lee Box. This .45/70 McKeever would have served the Remington Lees in use in the Navy until their replacement by the M1895 6mm Lee "Straight Pull" Rifle.

Photo 1.

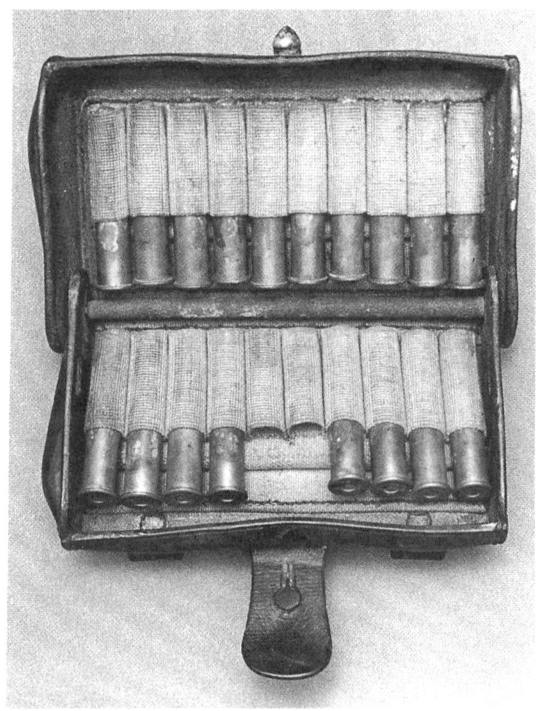

Photo 2.

"U.S.M.C." McKeever Cartridge Box for 6mm Lee Rifle Clips

Calibre: 6mm Remington Lee

Capacity: Four five-round clips; total–20 rounds.

Measurements:

OAL–	6 7/8"
OAW–	4"

Colour/Materials/Construction: When closed, this McKeever looks very much like any other Third Pattern box, except for the vertical, parallel lines of stitching on the back and face. This stitching secures the inner leather retaining straps, each of which is 1 3/16" wide and 6 1/4" long.

The box is unlined and the interior is undyed, though the retaining straps are dyed a reddish black. The belt loops are stitched at the bottom and riveted at the top.

Markings: The exterior of the right inner box end is stamped, **"ROCK ISLAND ARSENAL"** over **"W.T.C."** (an inspector).

The U.S.M.C. oval on the box faces measures 3 9/16" x 1 9/16".

Comments: A very scarce and desirable box that is rarely seen as good as the one pictured. It appears that most were ill-stored after U.S.M.C. service and most specimens need restoration.

It is felt that this box was not intended for combat use, but was a guard/parade/shipboard item and combat carriage of the clipped 6mm Lee ammunition was in a 10-pocket, blue web and black leather belt.

Photo 1.

Photo 2.

Late Military Academy/School McKeever Cartridge Box (Dummy)

Calibre: Dummy box; not intended for cartridges.

Capacity: None

Measurements:

OAL–	8 5/8"
OAW–	4 3/16"
OAT–	1 9/16"

Colour/Materials/Construction: This box is unusually well-made, of good quality black leather. It could be either arsenal or commercially made.

The exterior is stitched over a wooden block and there is no provision for opening. The slotted buckles captive on the back are typical of the cadet type McKeever, and were designed to take a light, white canvas or linen shoulder strap. They measure 1 3/16" x 2 3/16".

The body of the box is made of two end pieces and one wrap-around body piece.

Markings: "SMA" in an oval embossed on the face of the box. It is unknown what this stands for.

Comments: Though all specimens seen are like new, the boxes appear to have been in storage for some years. The brass buckles, also, have an old style flashing finish, rather than a more contemporary, simple polish.

Photo 1.

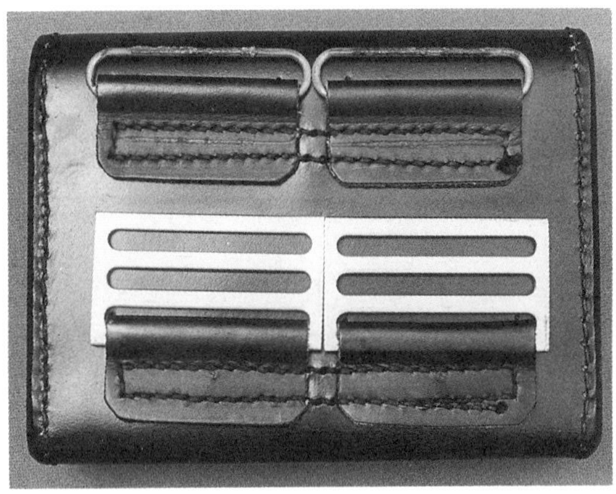

Photo 2.

Patented, Commercial and Experimental Cartridge Boxes

1867 Newcomb and Lyon Patent Cartridge Box

Calibre: .50/70 Government

Capacity: 30 rounds in three rows of ten cartridges each

Measurements:

OAH–	3 5/8"
OAW Flap–	7 15/16"
OAT–	2 3/4"

Colour/Materials/Construction: Now a brown leather, this box may have been black and the colour changed, as numerous accoutrements have, by virtue of the interaction of the dye and the particular leather used, i.e. the tanning process of hemlock, oak, etc.

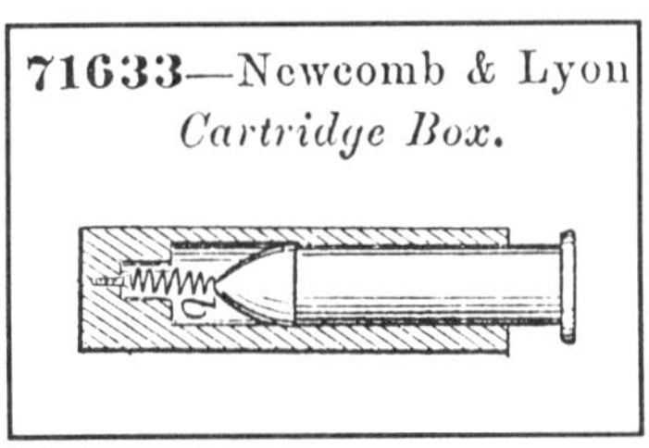

Photo 1.

This is a well-constructed box of commercial manufacture and post-CW design, i.e. no provision for shoulder sling. Two loops, each 15/16" wide, are stitched to the back of the box body for belt carriage only.

The cartridge holes in the soft wood block are fitted with the patented brass, spring-loaded plungers against which the bullet tip presses. This feature is seen on at least one other military box of the IW period and follows patent No. 71,633, awarded to Albert C. Newcomb and Benjamin Lyon of Springfield, Mass., in 1867. As shown in the patent drawing (Photo 1), a captive spring (or, as actually seen in production boxes, a spring-loaded piston) restrains the tip of the cartridge at the bottom of the round well bored vertically into a wooden cartridge block.

The cover flap is spring-loaded, having a leather-covered, sheet steel spring stitched to the back, underside of the cover flap and extending down into the box cavity. The obvious purpose of this was to keep the flap closed, except when pushed up by the hand. In practical effect, it is always in the way and impedes the grasping of the cartridge. A similar spring cover arrangement was patented by Augustus A. Bennett of Cincinnati, OH, 27 Jan. 1863.

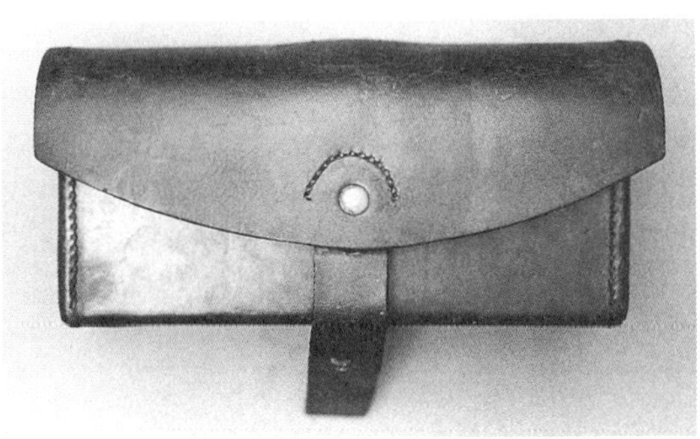

Photo 2.

Photo 3.

Markings: The only marking on the box is:

PATENTED
1867

stamped on the upper face of the box body.

Comments: The patent for this cartridge box reads as follows, ***"71,633.–ALBERT C. NEWCOMB and BENJAMIN LYON**, Springfield, Mass., assignors to themselves and SAMUEL W. PORTER, same place.–Cartridge Box.–December 3, 1867.–Explained by the claim.*

Claim.–A cartridge-box, having an elastic bed or support in the bottom of the chamber b, upon which the cartridge may rest while in place, substantially as herein described and for the purposes specified."

This is the only specimen of this box reported, and it may be unique. The Newcomb and Lyon design was an expensive and complicated solution to simple cartridge sticking. Simply boring the cartridge hole in the block a little larger solved the problem completely and inexpensively–almost certainly the reason this patent idea was not adopted by any large state militia or for federal use. The Newcomb and Lyon spring-loaded cartridge holder saw limited commercial production only one more time, then disappeared from the field of armament.

Photo 3.

Day's Patent Revolving Cartridge Box

Calibre: .50 Remington Rimfire

Capacity: (Estimated) 45 rounds in three rows of holes bored in a round wooden block.

Measurements: Unknown

Colour/Materials/Construction: The black leather box closely resembles a Civil War Cap Pouch, but has flat sides and a front flap with a center-mounted brass finial.

The flap pulls back to expose two sheet steel flaps, connected by wire stiffeners. This arrangement secures a heavy round wooden block that rotates on an axle, even with the closing flap finial. The outer surface of the block is grooved with three circumferential grooves, and those are each bored with staggered holes to admit the short, rimmed cartridges.

With the metal "ears" secured down, the round block could be rotated with the fingers to expose 4-6 cartridges at a time. When the box was closed and the flap secured, the cartridges in the box could not come loose in the box or drop out.

Markings: The front of the soft cover flap is marked in a stamped oval, **"DAY'S PATENT JAN'Y. 11. 70"**

Comments: Captain S. Allen Day of Bowling Green, Ohio, filed for a patent on a new concept for a cartridge box. Patent No. 98,748 was granted on Jan. 11, 1870. The synopsis of this patent, as it relates to the original patent drawing (Photo 1), reads as follows: *"Claim.–The block B, provided with the circumferential grooves, a, and having the radial holes or sockets b, arranged as herein described.*

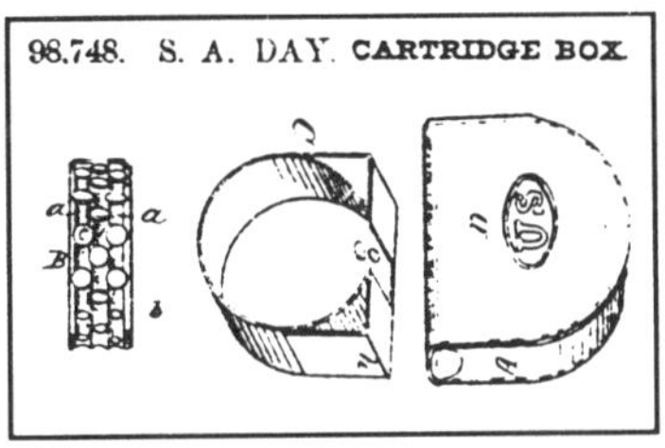

Photo 1.

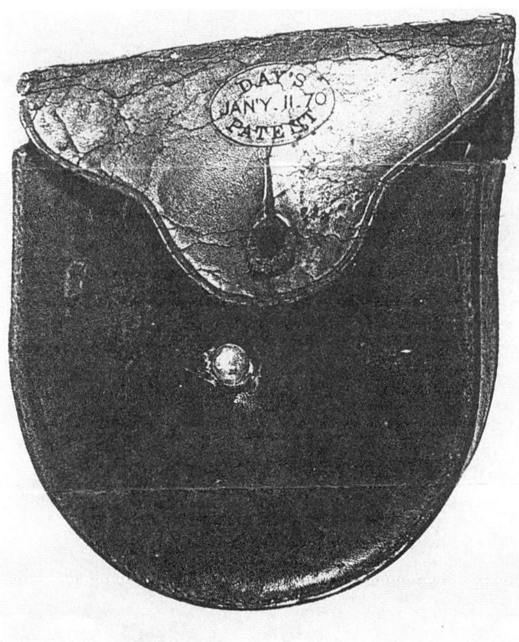

Photo 2.

Photo 3.

2. The combination, in a cartridge-box, of the block, B, internal case C, and external case A, all arranged substantially as described."

Several variations of Day's Cartridge Box were submitted for consideration to a board of officers meeting at Ft. Leavenworth, Kansas, in 1870. None of these boxes was found acceptable and the Army continued to search for more practical arms and accoutrements.

Day's concept was somewhat complex and, more importantly, was quite heavy, given its size–due to the wooden block that was central to its design. Today, specimens of the Day's Boxes are rare.

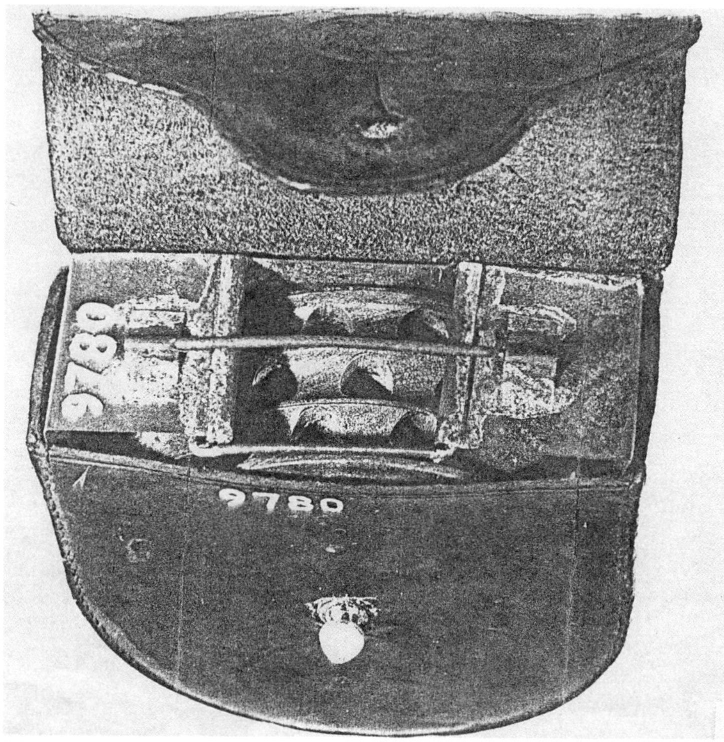

Photo 4.

Experimental "Cartridge Box for Colt, Smith & Wesson Schofield .45 Cal. Cartridge"

Photo 1.

Photo 2.

Photo 3.

Calibre: .45 Revolver (Schofield)

Capacity: 21 rounds in three rows of seven rounds each

Measurements:

OAL–	5"
OAW–	3 1/2"
OAT (Top)–	2"

Colour/Materials/Construction: This stiff, black bridle leather box has three rows of soft cloth loops hand sewn to stiffer cloth backing.

The box opens, clamshell-like, at the top and folds partially forward to expose the bases of the revolver cartridges. The box is restricted in its opening by leather strips, sewn on the inside of the box on each side.

Comments: The box was obviously one submitted to the Ordnance Department and, with the data discussed below, could be researched.

On a tag glued to the side of the box is a tag that reads, **"Cartridge box for Colt, Smith & Wesson Schofield .45 Cal. Cartridges. Ordnance office rec'd 3-11-1876"**. Under this tag is another glued tag reading, **"1336-76"** (an Ordnance Dept. file number.)

Photo 1 shows the box open, looking from the top down.

Photo 2 shows the box from the front, with the base of the closing tab.

Photo 3 shows the right side of the box, closed, with the belt loops on the reader's left.

The 1870 Roberts' Rifle Cartridge Box

Photo 1.

Photo 2.

Calibre: .50/70 Government

Capacity: 50 rounds in two levels of 25 rounds each.

Measurements:
OAL– 7 1/2"
OAH– 5 1/16"
OAT– 2 7/8"

Colour/Materials/Construction: This heavy bridle leather box is in the regulation black of the period. Manufactured at Watervliet Arsenal, it has many "regulation" design and construction features. The cover flap is secured by a leather tab to a bottom-mounted brass finial and has sewn-on, side-mounted "ears" that provide dust and precipitation protection for the upper portion of the box. (Photos 3 & 5)

When the cover flap is lifted, it exposes the upper layer of two nested cartridge boxes, each offering a layer of cartridges. The cartridges are held, bullet down, in three rows of holes cut in a heavy leather platform (two exterior rows of eight and a central row of nine). (Photo 4) The entire upper box is leather and is designed to tilt forward on a front mounted leather hinge (Photo 4) when its cartridges are exhausted. To prevent this from happening inadvertently, the back of the upper box has a long leather locking strap that protrudes from a slot in the back of the box (Photo 2) and buttons on the bottom-mounted finial, but under the cover tab. When the locking strap is unbuttoned,

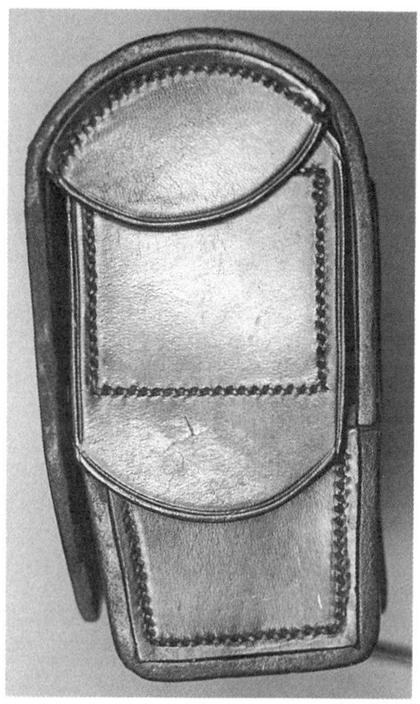

Photo 3.

pushing the upper "layer" or cartridge box forward exposes a second layer of 25 rounds underneath. These are held in a perforated leather platform exactly as the upper box. Note that the side panel of the upper box is cut with protective "ears" that provide the lower box of cartridges with dust and precipitation protection. (Photo 5)

The back of the box is fitted with two vertical belt loops (Photo 2) and there is no provision for sling carriage. It is noteworthy that such a large and complex box has only four brass rivets and one brass finial.

Markings: A 1" capital **"B"** is stamped over the marking

"U.S.
ARSENAL
WATERVLIET"

The **"B"** indicates the 1870 Board's identification of this box design.

Comments: 100 of these rare trial boxes were manufactured and few are known today. The box was submitted by Army Colonel Roberts and is the only pattern known to have been submitted by him at this time.

Photo 4.

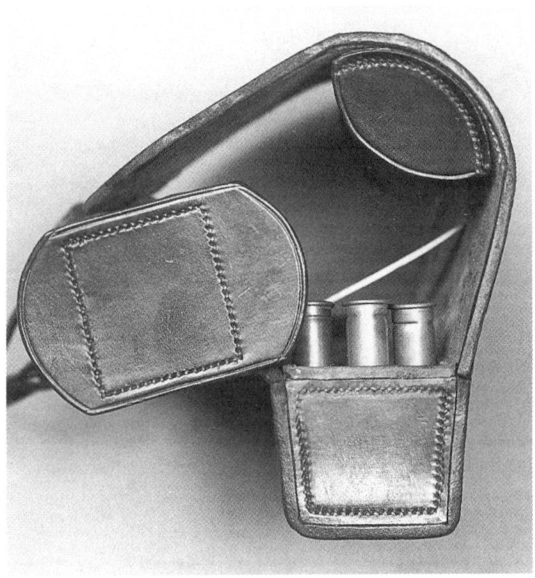

Photo 5.

Experimental 12-Round Revolving Pistol Cartridge Box

Calibre: .45 Army Revolver (.45 Colt)

Capacity: 12 rounds contained in two looped, rotating spindles

Measurements:

OAL –	4 1/4"
OAH–	2 5/8"
OAT–	1 7/8"
Height of Cartridge Loops–	1"

Colour/Materials/Construction: The body of this box is manufactured of stiff collar leather. The two rotating spindle shafts are wood, with six stiff (varnished) canvas loops attached to each one. The spindles are permanently affixed to the concave box body by vertical brass shafts. The closing tab is now broken off, so it is impossible to tell whether it had a brass escutcheon or not. The hinge for the folding box face was a leather strap, also now broken. Single belt loop on the reverse.

Markings: None but the embossed **"US"** in an oval on the cover.

Comments: The author, Dr. Scott E. Meadows, tentatively identifies this box as an A.D. Laidley patent, due to Laidley's submission of a **"revolving cartridge box"** to the 1874 Infantry Board.

As can be seen, this concept was complex and would have been expensive to manufacture. It, also, was not robust enough for the rigors of field service.

Photo 1.

Photo 2.

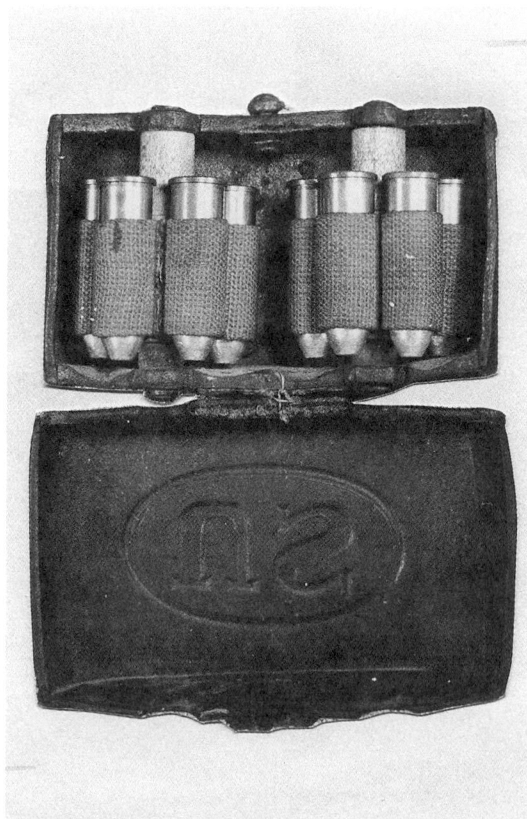

Photo 3.

Experimental 21-Round Revolving Pistol Cartridge Box

Calibre: .45 Army Revolver (.45 Colt)

Capacity: 21 rounds contained in three looped, rotating spindles

Measurements:

OAL–	7 1/4"
OAH–	3 1/2"
OAT–	2 1/4"
Height of Cartridge Loops–	1 9/16"

Colour/Materials/Construction: The body of this box is manufactured of stiff collar leather. The three rotating spindle shafts are wood, with six stiff (varnished) canvas loops attached to each one. The spindles are permanently affixed to the concave body by vertical brass shafts. The closing tab is plain leather. The hinge is a leather strap.

Comments: The author, Dr. Scott E. Meadows, tentatively identifies this box as an A.D. Laidley patent, due to Laidley's submission of a **"revolving cartridge box"** to the 1874 Infantry Board.

As can be seen, this concept was complex and would have been expensive to manufacture. It also was not robust enough for the rigots of field service.

Photo 1.

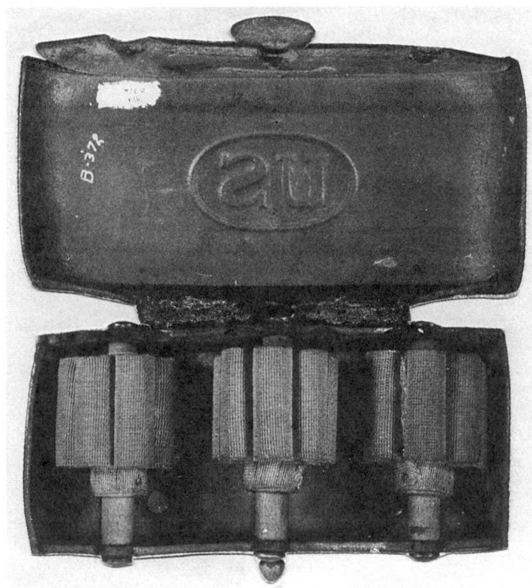

Photo 2.

Butler .50 Cal. Pistol Cartridge Box
Wooden Block Version

Calibre: .50 Cal. Remington Rolling Block Pistol M1871

Capacity: 19 rounds, contained in two rows of holes bored in a curved wooden block.

Measurements:

OAL–	7 7/8"
OAH–	2 1/2"
OAT–	2 3/8"

Colour/Materials/Construction: The body of this box is made of black bridle leather. The cover and body are mounted with the Carpenter Turn Key catch, in brass. The interior of the box is a bored wooden block. The cover is rather elaborately cut, overall, and the curved rear edge, riveted in two places, serves as the hinge for the cover. A central "ear" or scallop of leather on the back of the cover serves as a dust and rain guard.

The box is one of two types of Butler Box, the other having a tinned iron liner, fitted with 19 holes in two tinned metal plates.

Photo 1.

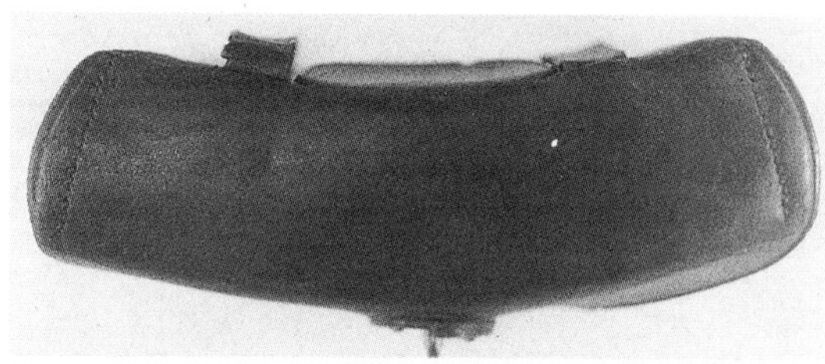

Photo 2.

Photo 3.

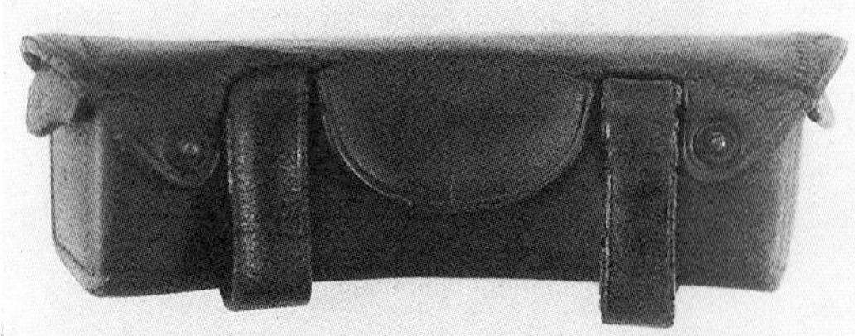

Photo 4.

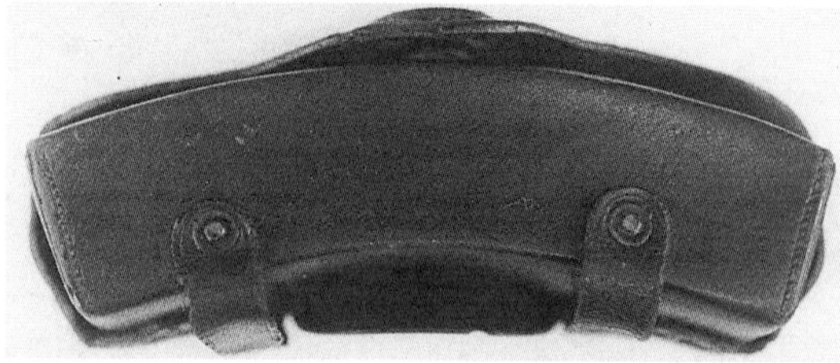

Photo 5.

Butler .50 Cal. Pistol Cartridge Box Tinned Steel Liner Version

Calibre: .50 Cal. Remington Rolling Block Pistol

Capacity: 19 rounds contained in two rows of holes in upper and lower metal sheets (Photo 1)

Measurements:

OAL–	7 7/8"
OAH–	2 1/2"
OAT–	2 3/8"

Colour/Materials/Construction: The body of this box is made of black bridle leather. The cover and body are mounted with the Carpenter Turn Key catch, in brass. The interior of the box consists of a tinned sheet metal box, in which the top and bottom are tinned sheets of steel, perforated with holes. The holes in each sheet are aligned so as to hold the rounds, bases up, in a vertical plane (Photo 1). The leather cover of the box is rather elaborately cut, overall, and the curved rear edge, riveted in two places in the back, serves as the hinge for the cover. A central "ear" or scallop of leather on the back of the cover serves as a dust and rain guard.

The box is one of two types of Butler Box, the other having a curved, bored wooden block with 19 holes in it.

Markings: It is understood that all Butler Boxes were stamped with a capital **"F"**, pursuant to instructions of the Chief of Ordnance, dated 24 Dec. 1873. Experimental cartridge boxes (pouches and a "cartridge vest") marked, A through H, including the Butler Boxes, were submitted to the 1874 Cavalry Board. The boxes were manufactured at Watervliet Arsenal, but it is unknown whether they were marked with the arsenal name.

Comments: The design of the Butler Boxes is attributed to Lt. John G. Butler, an Army Ordnance officer by Dr. Scott E. Meadows, author of U.S. Military Holsters and Pistol Cartridge Boxes. Though 5,000 of the Remington M1871 Army Pistol were purchased, only 100 each of the Butler Boxes (i.e. those with wooden inserts and those with metal liners) were manufactured by Watervliet Arsenal. These were distributed to Leavenworth Arsenal, Omaha Depot, San Antonio Arsenal and Benecia Arsenal. They were not particularly popular in field trial and did not see further manufacture.

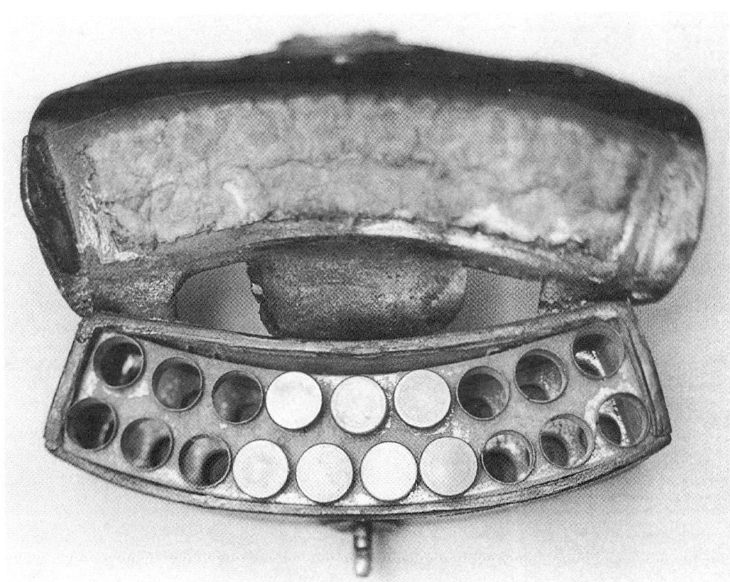

Photo 1.

"MVM" State Militia Cartridge Box

Calibre: .45/70 Government

Capacity :16 rounds in two rows of eight rounds each

Measurements:

OAL–	8"
OAH–	3 7/8"
Strap Height Above Box Body–	4 3/4"

Colour/Materials/Construction: his unusual box follows no particular pattern, and probably was the design of a state militia soldier or officer. The box is black leather with a tarred, separately-applied cover sewn to the body section.

The box fittings are reminiscent of the King's Patented Box with its three brass bottom finials (possibly for optional shoulder strap attachment), belt loops that extended under the box body and the securing cross strap that holds a portion of a strap.

Redundantly, there are two japanned iron shoulder strap buttons attached to the back of the box (Photo 2)–one of which holds a portion of a folded-construction leather strap.

The two rows of leather loops are simply sewn to the box body as a strip of leather.

The large, hollow brass volunteer militia plate is attached to the box face with seven small wires.

Markings: None, except the **"MVM"**, which could be a local or state volunteer militia.

Comments: This cartridge box is typical of commercially-designed and produced post-Civil War militia boxes that were not well designed nor, in many cases, robust enough for serious military use. The design of this piece incorporates a separately-applied tarred cover (a holdover from a much earlier periodr), low ammunition capacity and use of verdigris-producing leather loops. The redundancy of brass finials and iron buttons for shoulder strap attachment is also typical of the post-Civil War period.

This is an interesting and scarce militia cartridge box, whose exact origin is unknown.

Photo 1.

Photo 2.

Photo 3.

Post-Civil War National Guard Cartridge Box

Calibre: .50/70 Government

Capacity: 16 rounds in two rows of eight rounds each.

Measurements:

OAL–	6 3/8"
OAH–	3 5/8"

Colour/Materials/Construction: A black leather box of a design dating from well before the Civil War. It has an implement pouch on the face of the box body, though it lacks an inner flap. It has two belt loops on the back and has a bottom-mounted, brass finial for the closing tab. The closing tab itself is secured by a rivet and the curved stitching adopted in 1863. These factors, as well as the lack of bottom (Civil War-style) buckles, leads the author to believe this is a post-War modification of some type of CW rifleman's box. It is possible that, originally, an inner flap was present.

Currently, a 2 1/4" wooden block sits 1 1/2" down in the box cavity. On the face of the box, a stamped brass script **"NG"** is attached with four wires.

As can be seen from the back view, the back and front of the box are one piece of leather. The face or cover of the box is a separate piece of leather, attached just at the top of the loops on the back. The loops, incidentally, are stitched and riveted on both sides.

Markings: Faintly showing on the upper face of the box body is the early manufacturer's name: **"C.S. STORMS"**. This firm is not believed to have manufactured any such accoutrements after the War.

Comments: The author has seen a few of these boxes, and all show considerable wear and age. It is my belief that they were Civil War surplus, converted after the War, to make an inexpensive militia cartridge box for the 1866-1870 series of rifle muskets and rifles. Conversion would have consisted of removing one or two tins, replacing them with the wooden block, and removing the inner flap. The 1 11/16" x 1 1/2" sheet brass **"NG"** may have been separately procured by one or more states.

A scarce, desirable and early militia cartridge box, but not one often seen in excellent condition.

Photo 1.

Photo 2.

Photo 3.

New York State Militia Rifle Cartridge Box

Calibre: .50/70 Government

Capacity: 20 rounds in two rows of 10 cartridges each

Measurements:

OAL–	7 13/16"
OAH–	4 1/2"
OAT–	1 3/4"
OAW Loop on Back–	2 1/4"

Colour/Materials/Construction: Black leather box with full flap, reminiscent of pre-CW design. The soft wood block is bored with two lines of 10 holes each. In the bottom of each hole is a brass, spring-loaded device that operates as a spring-loaded plunger, against which the bullet tip presses. (Photo 3) This feature was patented by A.C. Newcomb and B. Lyon in 1867 and appears in an early Indian War period commercial cartridge box shown earlier in this section.

There is no provision for sling attachment. The single tab on the flap is attached with stitching and a single iron "flower" rivet of post-CW type.

Markings: No markings on any components except 1 1/4" high stamped brass **"M"** affixed to the center of the flap.

Comments: All of the boxes seen appear to have had little use, but do show issue. They are reputed to have been made at Sing Sing Prison in New York State. They are not common.

Photo 1.

Photo 2.

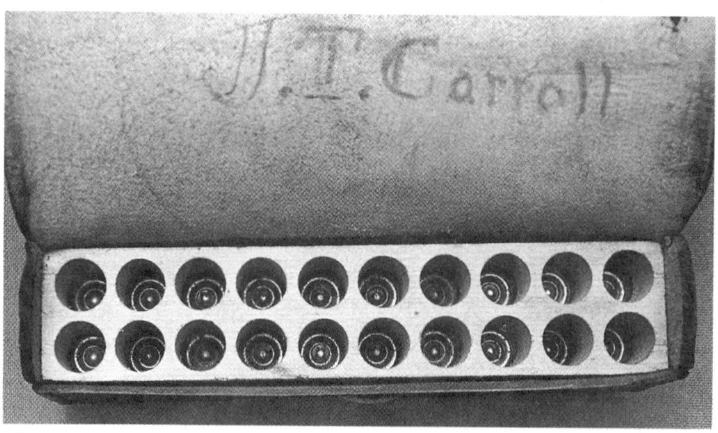

Photo 3.

Hunting or Forager Bag

Measurements:

OAL Bag–	7 1/4"
OAH Bag–	9 1/2"

Comments: A black, all-leather bag and shoulder strap set with two rows of leather loops for .45/70 cartridges (Photo 2) and a deep, inner cavity felt to have been intended for shot shells, such as used in the Army forager shot guns.

The styling of this bag and strap has a military flavor with its brass closing finial, large cover flap and 1874-type gusseted sides. The quality of stitching is excellent on the bag itself, but the two rows of .45/70 size loops are of field quality.

The straps are well finished, edged and are attached with brass rivets.

The upper inside edge of the bag has a 1/2" strip of brown wool encircling the mouth. It is stitched in on the upper edge.

Though the bag is unmarked, it is of arsenal or commercial quality. Only the rifle cartridge loops are added. It is a possibility that this is an arsenal-made hunting pouch, issued with forager shot guns. The added cartridge loops would make great sense for a team of foragers who wanted to be able to bring down birds or larger game–as well as having longer range Indian protection.

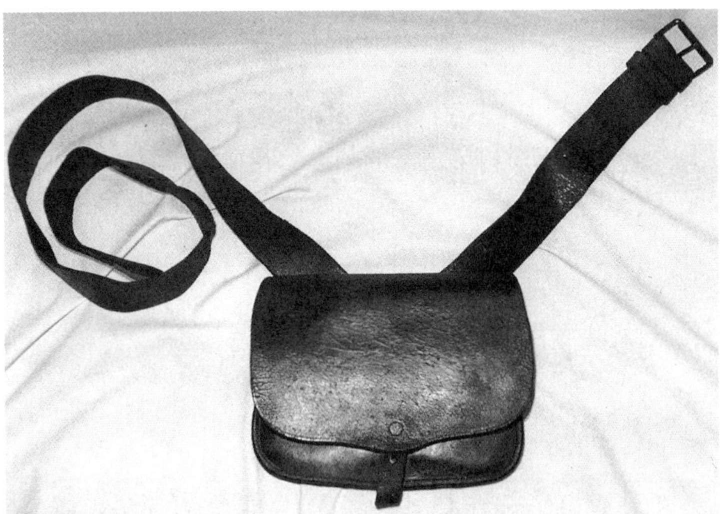

Photo 1.

Photo 2.

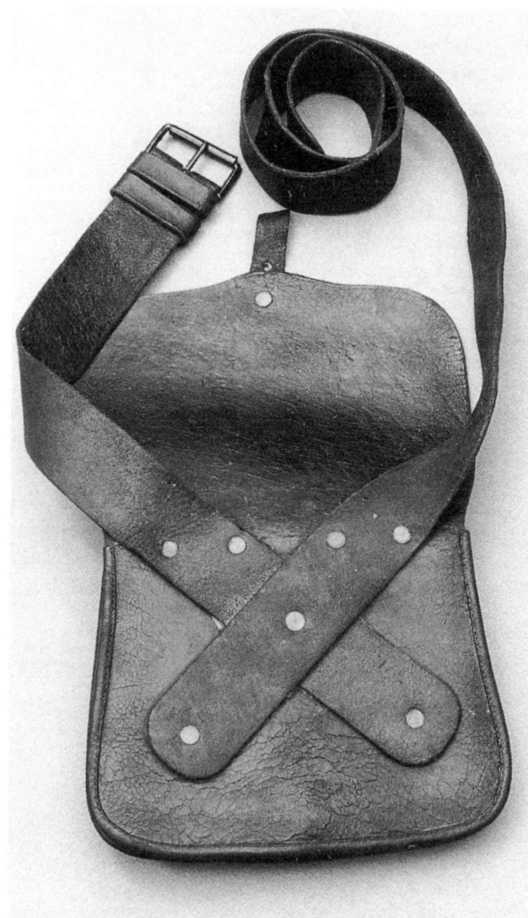

Photo 3.

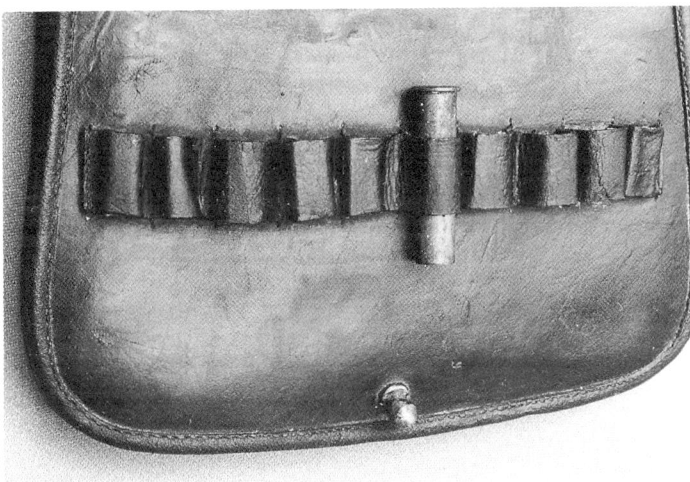

Photo 4.

Mann's Cavalry Carbine Cartridge Box M1864 (Large)-Converted

Calibre: .45/70 Government

Capacity: 32 rounds in four rows of eight cartridges each; 16 top, 16 bottom.

Measurements:

OAL–	6 3/8"
OAH–	5 7/8"
OAT–	2 3/8"

Colour/Materials/Construction: The Mann's Patent (Civil War) Carbine Cartridge Box was designed by William Dalton Mann of Detroit, Michigan. Commissioned as a captain with the 1st Michigan Cavalry in August 1861, Mann rose to the rank of Colonel in the 7th Michigan Cavalry by 1862. His interest in accoutrement design (and sale) was such that he resigned in 1864 and set about full time design and promotion of a unique series of infantry and cavalry cartridge boxes. Five of his cartridge boxes are shown in Vol. I of Frederick P. Todd's American Military Equipage 1851-1872. Though never adopted as regulation by the federal government, his boxes did see use and were manufactured by both C.S. Storms, NY, and Gaylord, Chicopee, Mass.

The box shown, though discussed in Todd, is not pictured there and differs from the displayed Mann boxes by displaying two vertical belt loops and a connecting horizontal strap on the back. (Photo 2) These are original to this M1864 Cavalry Box (Large) and were not added during the conversion process.

Conversion of this box took place after 1873 and was accomplished by removal of the tinned steel liner and its replacement with a soft wood block, (Photos 4 & 5) bored on both ends with 16 holes in two rows–for a total of 32 holes. A grained, russet, soft leather flap was both stapled and iron nailed to one side of the block. This flap pulled over the lower two rows of cartridges and was secured on the block face by buttoning over a low brass finial. The purpose, of course, was to allow the block to be reversed for use of the lower 16 rounds when the upper 16 had been used. One further modification of the original Mann Box was the removal of the long, centrally-mounted shoulder suspension strap from the back of the box by removing a central rivet and two rows of stitching. The entire conversion was practical and cost-efficient.

Markings: The large US logo on the large flap of the box is marked: **"COL. MANN'S PATENT REISSUED JUNE**

Photo 1.

Photo 2.

7TH 1864–E. GAYLORD MAKER CHICOPEE, MASS."

To the right front of the large flap is an inspector's cartouche (**"Shepard"**) over an Old English script : **"U.S."**

Comments: This box was originally manufactured by a commercial manufacturer under federal contract. It may have been retained in federal inventory and converted in an arsenal, but is much more likely that it was either sold in its original Civil War configuration to an entrepreneur who converted it for resale, or was sold to a state and was then converted by a commercial concern for state militia use.

This is the only conversion of the M1864 Mann Cavalry Box known to the author.

Photo 3.

Photo 4.

Photo 5.

King's Patent Cartridge Boxes–Overview

In 1870, the year that Josias King patented his folding cartridge box, the U.S. Army was still some years from the adoption of field-worthy and efficient cartridge boxes. What was then in service was a rather mixed collection of modified Civil War cartridge boxes–some with wooden blocks, some with fleece lining and none completely satisfactory. In this perspective, the relatively compact King's concept was, indeed, a new idea.

As seen in the original patent drawings of 15 Dec. 1870, King's cartridge box was a simple, fold-open box, secured at the front and sides with leather tabs fitting over brass finials. Specifically, King's patent was as follows: ***"109,365.–CARTRIDGE BOX.****–Josias R. King, Washington, D.C. assignor to William M. Tileston, same place. * Claim.–1. The plate A, whether made in one or more pieces, and provided with rows of loops for holding cartridges, substantially as described.*

2. The plate A, provided with rows a b of loops on one side, and a row or rows, d, on the other side, so arranged as to fold compactly together and be covered by the plate, substantially as specified."

An article in the 26 August, 1871, Army and Navy Journal indicates that a belt-mounted, 30-round box was then being tested by the U.S. Army. It went on to note that this box design had been submitted to the New York State (Militia) Board and that an informal meeting of Connecticut military (militia) officers had given it "general approval." What field service the King's Box may have had, oddly enough, probably came in Russia where, apparently, a regiment in the "practice camp" was issued the box. These were, doubtless, made in the U.S. on a limited contract.

After that time, little was heard of the King's Box–and for good reasons. First, the box was needlessly complex. It usually carried 24 to 30 rounds, and had to be unfolded for the user to reach 1/3 of his ammunition. A simple box of three rows of cartridges, all with their bases up and available to the wearer would have been faster to use, cheaper to manufacture, more durable in field service and could offer more protection to the copper cartridges. Second, as alluded, the design required light leather (so that it could be folded); light, folding side pieces or ends, and two additional light side tabs. Along every fold, the leather was scored to ease folding. All of this made for a box that had easily torn parts. Third, both the early (1871) box reviewed in the Journal and the most commonly-seen .45/70 King's boxes both use leather loops. Copper cartridges in these boxes, contacting the leather of the loops as well as the box body, form verdigris readily–a serious drawback that plagued the Army for years to come.

What the King's Box did offer was a compact method of carrying a number of cartridges arranged in loops for relative easy access. It was an idea that was quickly eclipsed by more sturdy and cost-efficient designs, but it was one of the very first of the post-War boxes to show new thinking on the subject.

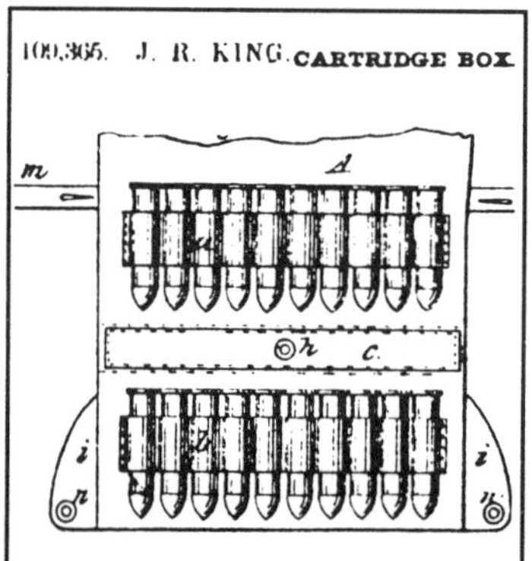

Photo 1.

King's Patent Cartridge Box–Possible First Pattern

Calibre: .50/70 Government

Capacity: 42 rounds in three rows of 14 rounds each

Measurements:

Closed (Photo 1)	
OAL Cover–	10 3/4"
OAH–	3 1/2"
OAT–	2 1/2"
Fully Open (Photo 3)	
OAL with Tabs–	13 3/4"
OAH with Tabs–	14 1/2"

Colour/Materials/Construction: Besides the obvious size and cartridge capacity of this box, its most striking feature is its styling–following the CW rifle musket cartridge box in closure method, vertical belt loops and, most significantly, its continuation of the bottom-mounted buckles for attachment of a Civil War-style shoulder strap. This latter feature is only found in this pattern King's Box.

The black leather used in the box body is a hard finish collar leather, skived (or scored) on the inside of the folding side panels. To give greater stiffness, the full interior of the box is lined with a dark blue striped ticking, much like early pillow or mattress ticking.

Each row of loops is formed of a strip of multi-coloured webbing 1 1/2" wide, sewn to the ticking and leather box body to form loops. The colours on each loop, from top to bottom, are red, navy blue and white–with the sequence repeated in reverse.

As noted in the introductory comments, the supposed advantage of the King's Box was its safe carriage of ammunition. Upon opening the cover, this box presented the wearer with 28 rounds, base up, and 14 rounds, base down. After exhausting the first 28 rounds, the two side tabs were unbuttoned and the face of the box fell forward, exposing another row of cartridges, base up. (Photo 3)

Photo 1 shows the face of the closed King's Box. Photo 2 shows the box open, exposing one row of cartridge loops stitched on the front of the box body and another row of base-up cartridges inside the box but not seen in the photo. Photo 3 shows the front of the box fully opened. Photo 4 shows the back of the box with horizontal sling guides, belt loops, two sling buckles and an inverted row of cartridge loops. (These would have been emptied before the box would have been opened under normal circumstances.)

Markings: None.

Comments: The lack of any markings on this box, its calibre and its early styling (differing from the box reviewed in the Army and Navy Journal of August 1871) lead the author to think this may have been a pre-patent or very early trial of this idea. In any event, it is unlike any other reported King's Box.

Photo 1.

Photo 2.

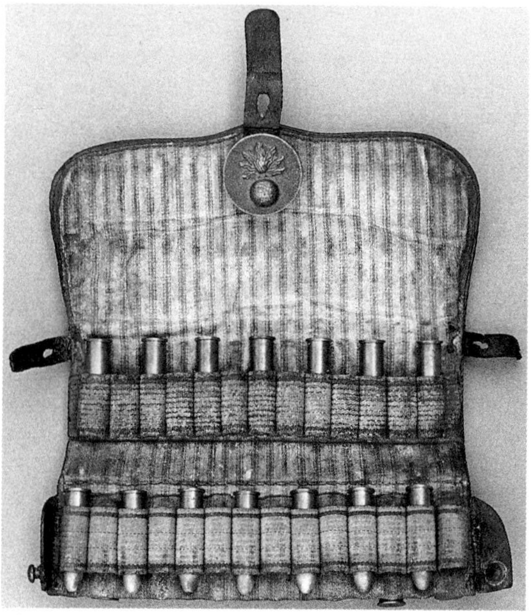

Photo 3.

Photo 4.

King's Patent Cartridge Box–Later Pattern

Calibre: .45/70 Government

Capacity: 30 rounds in three rows of 10 rounds each

Measurements:

Closed (Photo 1)	
OAL–	7 3/8"
OAH–	4"
OAT–	1 7/8"
Fully Open (Photos 3 & 4)	
OAL with Tabs–	12 3/4"
OAH with Tab–	14 3/8"

Colour/Materials/Construction: This black leather box is fitted with brass finials and rivets. All leather is fairly soft collar leather and, where stiffness is required, that is achieved by a doubling of the thickness of the leather. The sharp, interior bends of the leather (box bottom and side flaps) are skived to allow a sharp fold. This, however, seriously weakened the leather along these bends.

Cartridge loops are very soft leather, somewhat lighter than that of the box body.

Basic design of the box is very much a follow-on to the common rifle musket cartridge boxes of the Civil War. The two outside finials on the bottom of the box were to affix the lower ends of a shoulder strap, which then would pass through the horizontal securing strap on the upper back of the box. Two angled belt loops were also available on the back of the box.

As can be seen from Photos 1 and 4, unbuttoning the cover of the box offered two rows of cartridges, base up, to the wearer. By further unbuttoning two side tabs from their finials, the face of the box body unfolded downward, exposing a previously-inner row of cartridges, base up. Photo 5 shows a comparison of sizes between the later, more common .45/70 Cal. King's Box (top) with the First Pattern Box (bottom).

Markings: All examined specimens of this particular variation of the King's Box have two large stamps on the back of the box.

Photo 1.

The first reads, **"Agent–Shuyler, Hartley and Graham New York."** The other large stamp reads, **"Josias R. King's Pat. Nov. 15 1870. W.M. Tileson, Assignee."** No other marks appear.

Comments: The comments made on the earlier pattern King's boxes apply to this pattern. This variation, however, does appear to be the last venture on any market and none of these boxes shows use by even state militias–even though it was obviously made with military usage in mind.

A black, .58 calibre King's Box in this more standard pattern is known and is clearly stamped **".58."**

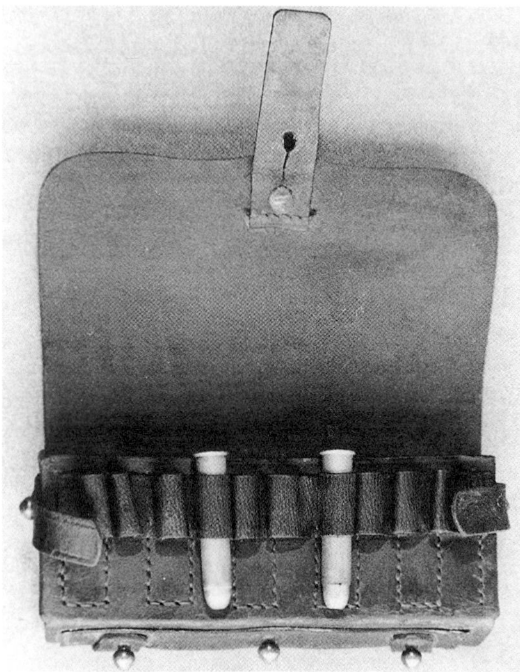

Photo 2.

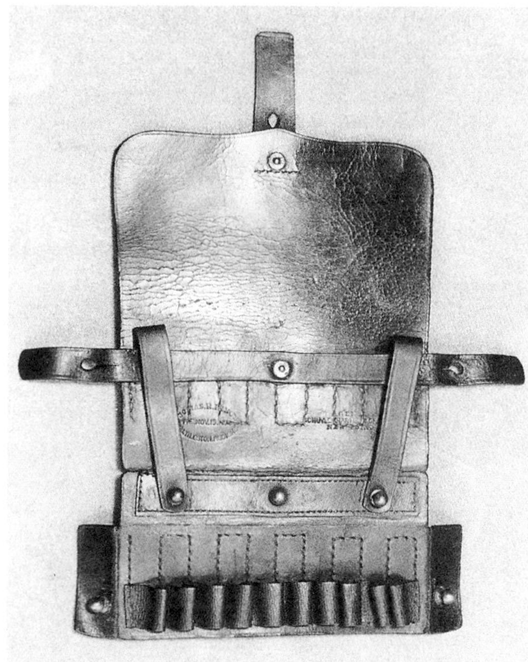

Photo 3.

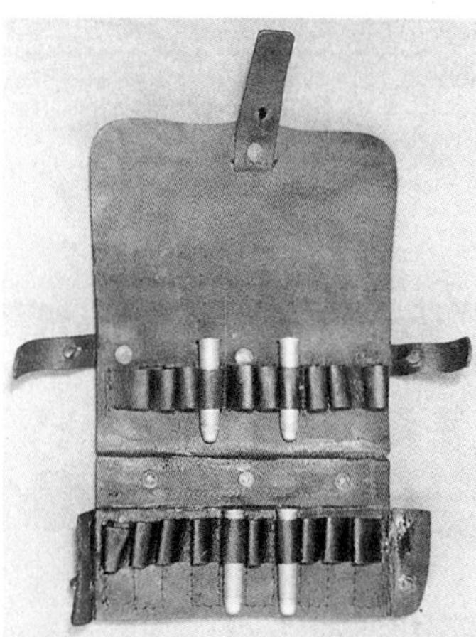

Photo 4.

Photo 5.

King's Patent Cartridge Box-20 Gauge Shotshell

Calibre: 20 Gauge Shotshell

Capacity: 30 rounds in three rows of 10 rounds each

Measurements:

Closed OAL–	10 1/2"
OAH–	4 1/2"
OAT–	3"
Fully Open (Photos 2 & 3)	
OAW without Tabs	12 3/4"
OAH without Tabs	13 1/2"

Colour/Materials/Construction: This box appears to have originally had a light tan colour and was fitted with the standard King's Box brass fittings. The belt loops on the back are standard for the pattern, but they continue on to beome an integral shoulder strap–attached by brass rivets. The strap is adjustable and has a commercial type horseshoe buckle. The straps could have been much more strongly attached by adding stitching to the area of attachment by rivets, but this weakness is consistent with King's other design flaws.

Markings: These are stamped on the bottom of the box (Photo 4) and contain the standard King's patent and Schuyler, Hartley and Graham information.

Photo 1.

Photo 2.

Comments: The calibre of this box offers the tempting possibility that King's intended market for this box was the military, for its forage guns of the period. Though the "Trapdoor" Forager came out in 1881, other shot guns are known to have been purchased by the Army, e.g. the Remington Rolling Block Forager in 20 gauge. It is not impossible that this cartridge box was a last attempt to interest the military as well as the private hunting market.

Photo 3.

Photo 4.

Frazier's Cartridge Boxes—Overview

On Jan. 5, 1878, Jos. W. Frazier of New York City, filed for a patent, having "invented a certain Improvement in Cartridge Pouches." Patent 202,642 was granted him April 23, 1878.

Photo 1 shows the original Frazier patent drawing. Relating to that drawing, the Frazier patent is described as follows: *"My improvements relate to that kind of pouch which is intended to receive and contain cartridges in packages open at one end; and my invention consists, first, in a box, open at the top, of proper size to just receive a package of cartridges, and having a hollow cover, the inner side of which is a spring-strip which in the act of closing the cover, is brought to bear upon the ends of the cartridges, and which yields, to allow the cover to be pressed down and buttoned, and then reacts upon the cover, pushing it up and tightening the button-loop, and at the same time holding all the cartridges firmly down, so that they cannot move about and be shaken by the movement of the wearer of the pouch."*

"The second feature of my invention consists in combining the hollow cover, provided with the yielding interior cushion, with a shield or flap which overlaps the front side of the pouch, and is hinged to the bottom thereof. The cover and shield, by reason of being hinged to the bottom of the pouch, drop completely out of the way when the pouch is opened, and the shield affords a smooth surface of ample space, upon which any letters or figures can be conveniently affixed."

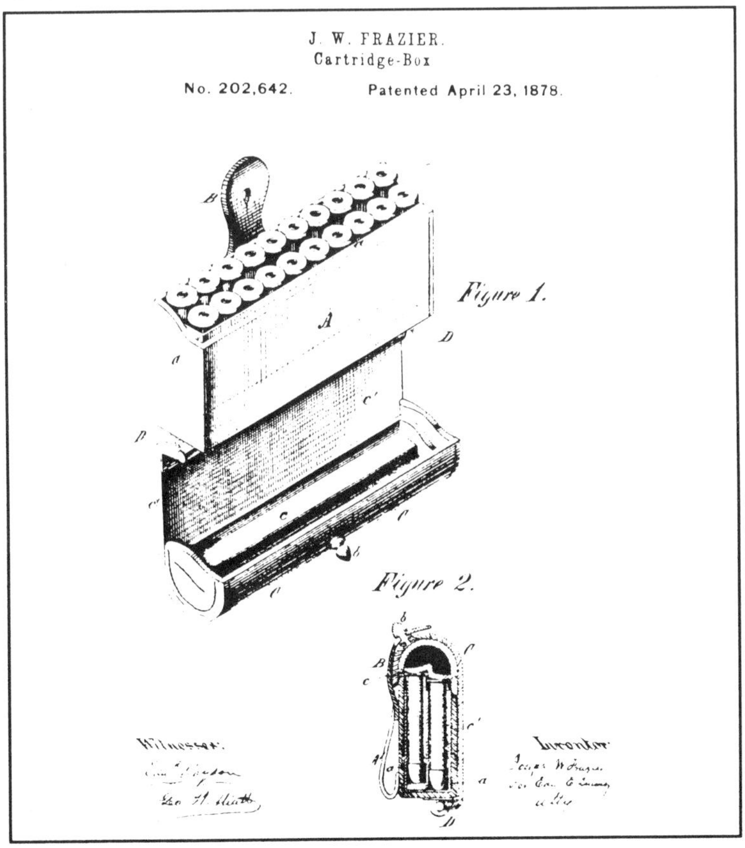

As the reader can see, the early Frazier Cartridge Box we all know incorporates only the folding cover feature from the original patent. Between the patent and the actual boxes manufactured, utilization of a single cavity for a package of cartridges gave way to the incorporation of a bored wooden block.

Though the "yielding interior cushion" concept did not make the transition to production, the "spring-loaded strip" on the underside of the cover did. Interestingly, a Frazier-type spring-strip was later used in production specimens of the Hitchcock Cartridge Box of 1879—though this feature is intended as a lid-stiffener, not a cartridge restraint.

Frazier's interest in cartridge box design continued for another decade and, in 1887, he was granted three patents on concepts relating to cartridge boxes with vertically-sliding covers—as well as other features. None of these patents was accompanied by a model, but a cartridge box incorporating features from two or three of these 1887 patents was manufactured and is shown and discussed at the end of the Frazier Box section. It is probable that this was Frazier's last effort to interest the armament industry and the military in his designs.

Frazier's Cartridge Box–Belt Loop Pattern

Calibre: .50/70 Government

Capacity: 18 rounds in two rows of nine cartridges each.

Measurements:

OAL–	7 3/8"
OAW–	3 7/8"
OAT–	1 13/16"

Colour/Materials/Construction: A black, bridle leather box with brass fittings.

Like the Shoulder Strap Pattern Frazier, the cover/face of the box rounds over the top and upper back to form a full cover with a closing tab. The cover/face is hinged at the bottom on two brass hinges, riveted to the bottom of the box body. The cover/face pulls forward at the top, exposing the cartridge block, but–unlike the McKeever Box and the original Frazier patent–is restricted in movement and only opens about 55 degrees.

The round brass tab escutcheon is riveted to the tab with three rivets. The light wood cartridge block measures 2" x 9" and has a top sloping toward the front. A spring steel stiffener is riveted lengthwise on the inside of the box cover (to prevent crushing when open).

Both end caps of the cover/face, the closing tab and the belt loops all have flat, black-painted commercial iron rivets. The loops are 15/16" wide and are both stitched and riveted to the back of the body. A hollow, NG (National Guard) brass plate 2 3/16" x 1 1/2" is affixed to the face cover with two loops through the cover.

Markings: The front left face of the box body bears the following marks, about 1" high x 2" wide:

FRAZIER'S PATENT
FEBY 5TH 1872
REISSUED MCH. 15TH 1878
PATENT APRIL 23RD 1878

To the right of that mark is another of the same size:

RIDABOCK & CO.
SUCCESSORS TO
J.H. McKENNEY & CO.
NEW-YORK

Comment: This is the most common of the three known Frazier Box patterns and may have served several state infantry militias but was never adopted for federal service.

The Frazier is relatively uncomplicated, but is heavier and more fragile than the popular McKeever Box. In particular, when the cover/face of the Frazier is open, it is very subject to severe damage if put under the kind of torque occasioned by a soldier falling or rapidly assuming a prone position and/or rolling–all of which could happen in a combat situation.

Photo 1.

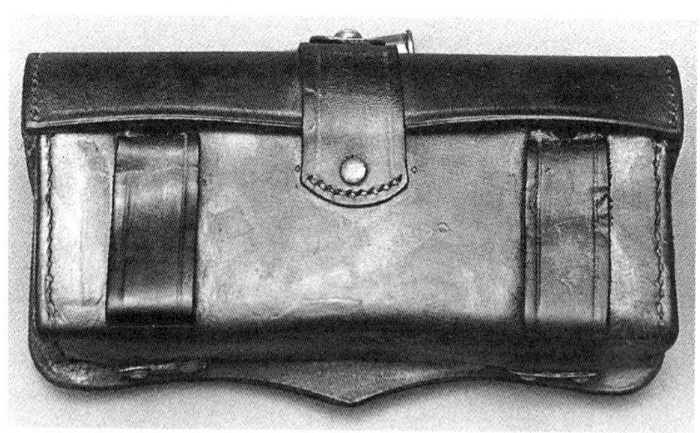

Photo 2.

Photo 3.

Frazier's Cartridge Box–Shoulder Strap Pattern

Calibre: .50/70 Government

Capacity :18 rounds in two rows of nine cartridges each

Measurements:
OAL– 7 3/8"
OAH– 3 7/8"
OAT– 1 7/8"

Colour/Materials/Construction: A black, bridle leather box with brass fittings.

Like the Belt Loop Frazier Box, the face of the box rounds over the top and upper back to form a full cover with a single closing tab. The cover/face is hinged at the bottom on two brass hinges riveted to the bottom of the cover/face as well as the bottom of the box body. The cover/face pulls forward at the top, exposing the cartridge block, but–unlike the McKeever Box and the original Frazier patent–is restricted in movement and opens only about 55 degrees.

The round brass tab escutcheon is riveted to the tab with 3 rivets. The light wood cartridge block measures 2" x 9" and has a top sloping toward the front. A spring steel stiffener is riveted lengthwise on the inside of the box cover (to prevent crushing when open.)

The shoulder strap version of the Frazier Box is clearly factory-made but it incorporates both ends of the narrow (1 7/16") IW carbine sling shown in the 1881/85 Ordnance Memo. No. 29. As seen in the photo, the left rear of the box body has an angled rivet and sewn-in leather chape holding a 1 7/8" wide flattened oval brass ring. A 47 1/2" long section of the (carbine) sling is sewn over this brass ring. On the right side of the back of the box body, a 9" leather section of the (carbine) sling is sewn and riveted, extending parallel to the axis of the box. This section of the sling holds the 2" x 2" brass frame buckle common to the carbine sling.

Both end caps of the cover/face, the closing tab and the attachment of the sling ends to the box body all have flat, black-painted commercial iron rivets. A hollow, **NG** (National Guard) brass plate, 2 3/16" x 1 1/2", is affixed to the face cover with two loops through the cover.

Markings: The tip end of the sling is marked with a standard **"ROCK ISLAND ARSENAL."** On the more flexed, shorter section, is the stamp "TROOP C", other illegible marks and "13". The front left face of the box body bears the following marks, about 1" high x 2" wide:

FRAZIER'S PATENT
FEBY 5TH 1872
REISSUED MCH. 15TH 1878
PATENT APRIL 23RD 1878

To the right of that mark is another of the same size,

RIDABOCK & CO.
SUCCESSORS TO
J.H. McKENNEY & CO.
NEW-YORK

Comment: It is felt that this pattern of the Frazier Box could have been used by either militia, cavalry or infantry, but it is much more likely that it was for Infantry only, since cavalry cartridge belts were more practical and easily available from the federal government. It was never adopted nor used in federal service.

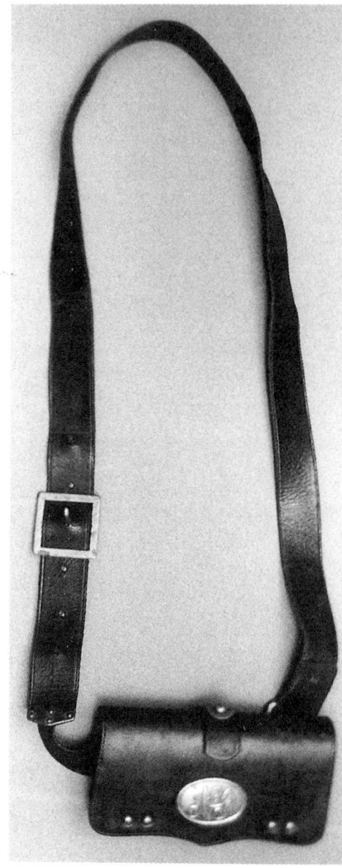

Photo 1.

It is felt that the carbine slings used in the construction of this pattern of Frazier were purchased used from government stores–hence the troop markings.

This pattern of Frazier box is uncommon.

Photo 2.

Frazier's Patented Sliding Cover Cartridge Box

Calibre: .45/70 Government

Capacity: 20 rounds in the standard military cardboard box (removable).

Measurements:

OAL–	7 3/8"
OAH–	3 1/2"
OAT–	2 1/4"

Colour/Materials/Construction: This black bridle leather cartridge box has two patented features that make it unique: 1. It has a vertically-sliding cover that drops below the box body once the box flap is opened. 2. It has a multi-purpose brass frame riveted to the back of the box body with five rivets.

By 1887, the date of Frazier's patents, new cartridge box designs for the services were predicated on their carrying the standard 20 round cardboard cartridge boxes sealed with paper labels and opened with a pull-string. This Frazier Box, as well as the patent concept boxes, used the drop-in 20 round cartridge boxes that could be carried sealed until such time as needed. Since the cardboard boxes had integral separators, boxes like Frazier's and Hitchcock's needed only a box-shaped hollow to accept the cardboard containers.

Frazier's patent #368,578 is lengthy and relates both to a vertically-sliding cover and a metal "frame" attached to the back of his cartridge box body. (Photo 1) His patent concept "frame" had four basic functions, which are listed below. However, note that no patent model was submitted and that, by the time a box was actually made up, the frame design had changed. In use, the patented frame served four functions:

1. The frame serves as a guide for the patented sliding cover feature of the box.

2. The curved ends of the plate, covered with black leather, retain two leather waist belt loops.

3. The frame serves as a stiffener for the box body. Note that many other designs use a bored wooden block as an internal stiffener or, as in the Navy's Hitchcock Box, use a sheet steel stiffener.

4. The long, C-shaped metal frame (Photo 4) serves as a standoff, to give space for ventilation between the back of the box and the body of the wearer.

The "frame", as it was actually manufactured, became a solid, sheet brass plate with rolled edges and curved ends that supported leather belt loops. (Photo 4) This later design was probably more cost efficient to produce.

Frazier's patent #368,579 describes a vertically-sliding cover

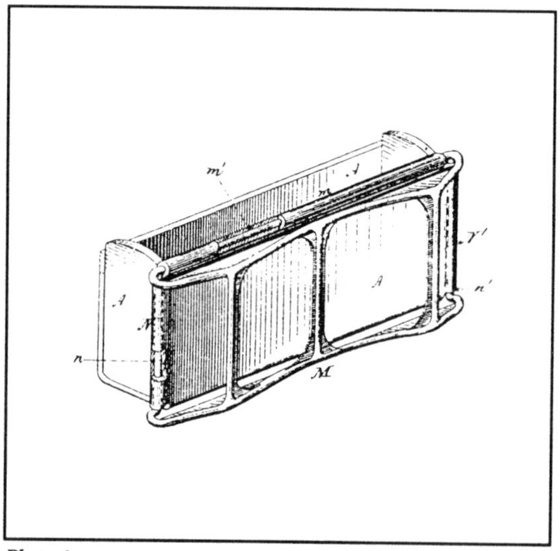

Photo 1.

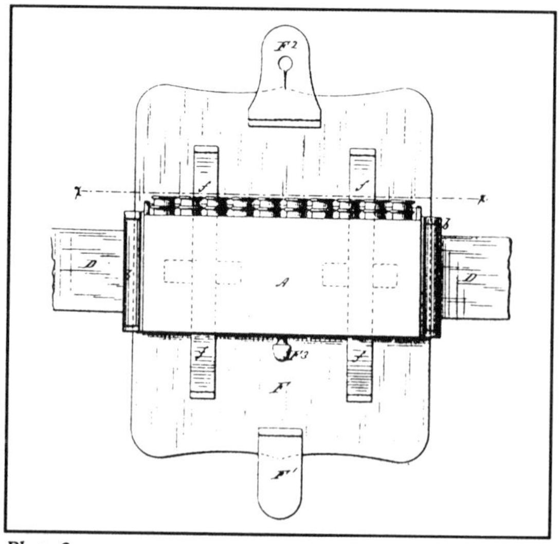

Photo 2.

on two internal leather guides. (Photo 2) As with the frame by the time the box was actually produced, the design had been simplified to one centrally-mounted internal cover guide strap.

Photo 3 shows the front of the actual cartridge box and Photo 4, the back. The large brass plate with rolled edges is covered with leather, but the edges can be seen in this photo. The ends of the plate curve back toward each other and each carries a leather belt loop.

Photo 5 shows the box, just opened, with the flap up. Photo 6 shows the flap pulled all the way down so as not to interfere with the taking of cartridges. Note that the bottom of the flap has an unperforated pull tab.

Markings: Most armament and accoutrement inventors of the time laid their greatest hope in interesting the military in their patented concepts. To this end, Frazier's new cartridge box was manufactured in regulation black leather and has a regulation "U.S." in an oval wet-pressed on the cover.

Inside, on the face of the box, is stamped, **"PATENTED AUGUST 23 1887."** No other markings are evident.

Comments: On 23 August 1887, Joseph W. Frazier of New York, NY, received three patents. All three were entitled "Cartridge Box", and all three related to sliding cover boxes. The first patent, #368,577, specifically addressed a "cartridge box provided with a cell or cells, each of which is open in front and has elastic side walls adapted for receiving and holding by friction a cartridge-magazine". Patent #368,578 dealt with both the sliding cover and the metal frame on the back of a Frazier Box and patent #368,579 dealt with the sliding cover feature as well as other minor construction details.

Frazier's cartridge box concepts are all interesting but, as actually manifested in the specimen box shown, are both complex and easily damaged. Frazier's heavy reliance on slender, stitched straps and stitched tabs could have been improved with additional rivets, but his designs offered no significant advantages over the sturdy McKeever. Further, the use of a large (2 1/4" x 7 1/4") brass plate on the back made this design less attractive in terms of cost when compared to the McKeever or the Hitchcock Cartridge Box. But, certainly, the overriding consideration working against the new Frazier Boxes and any other such designs, was simply that the day of the cartridge box as anything but a garrison item was over. The Army had plenty of McKeevers in stock for the light duty of guard mount and parade. For serious field use, the cartridge box was left in garrison and the web cartridge belt of Anson Mills and Henry Orndorff were loaded and carried.

The Frazier Box shown here is quite rare and was probably part of a very limited production. It probably was submitted to the Army but there is no documentation of that or possible, limited trial.

Photo 3.

Photo 4.

Photo 5.

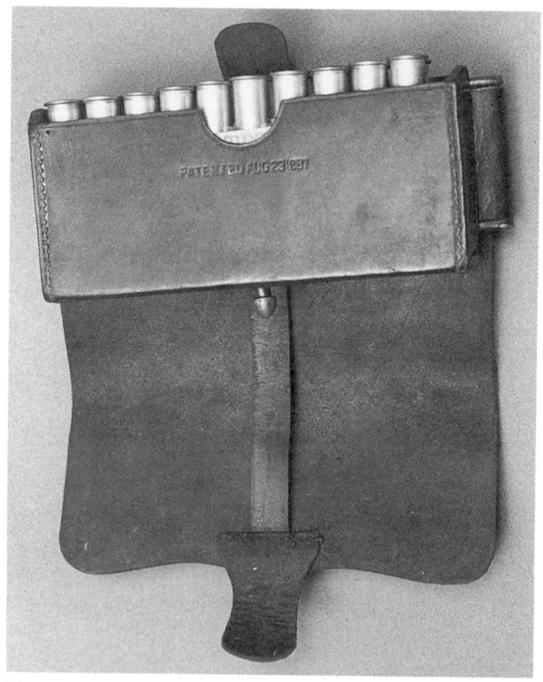

Photo 6.

Hitchcock's Patented U.S. Navy Rifle Cartridge Box

Calibre: .45/70 Government

Capacity: One standard, government issue cardboard cartridge box of 20 rounds.

Measurements:
OAL– 7 1/8"
OAH– 3 7/16"
OAT– 2 1/4"

Colour/Materials/Construction: This box is definitely brown, and that colour is felt to be due to a degradation of the black dye in a leather not tanned to strict government specifications. (Note: this was a commercially-manufactured box.) This problem occurred regularly in the history of U.S. leather accoutrements.

The design of this box is unique in that it has a finial on each end of the box body. The finial on the right end is mushroom-headed and holds that end of the cover captive. The finial on the left end is a standard, ta-

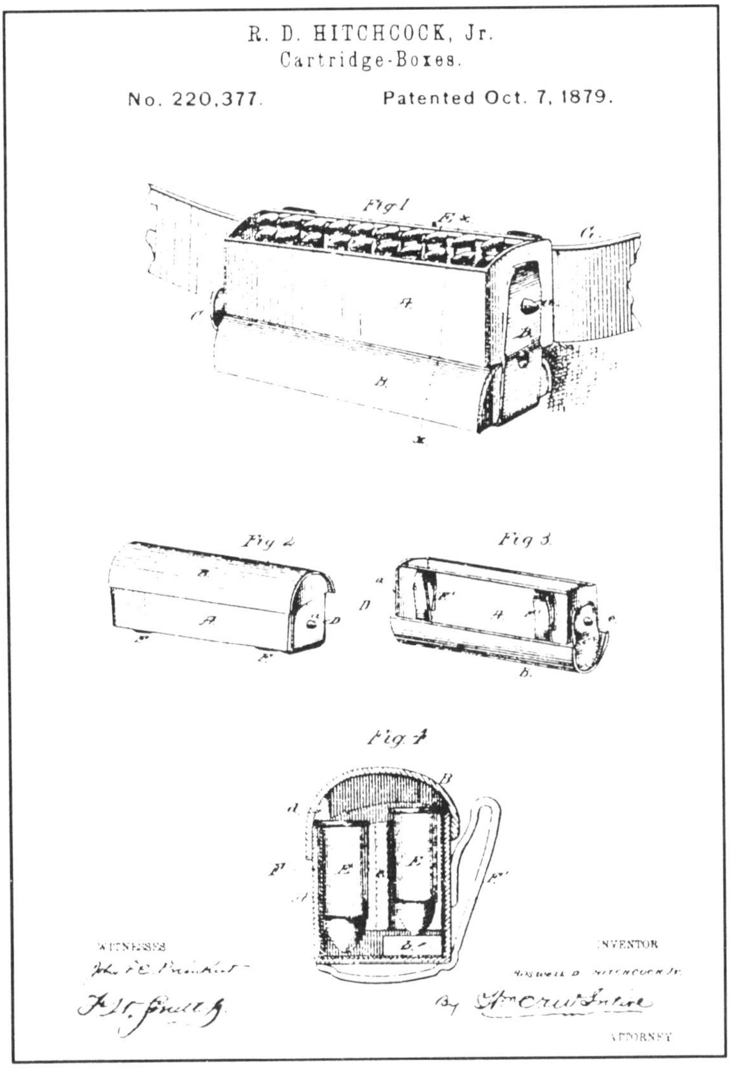

Photo 1.

pered type that allows the cover tab to be pulled from engagement and the entire cover pivots forward, around, then up around the base of the box. This fully exposes the cartridge box inside. On call, the pull string on the sealed cartridge box was pulled and cut the top of the cardboard cartridge box off, exposing 20 rounds.

Lengthwise, in the upper curve of the leather box cover, a leather-covered sheet steel stiffener is riveted in five places to the upper lip of the box body.

Markings: A large **"USN"** in an oval is embossed on the front and top of the movable cover. On the back of the box body, the following information is stamped in an oval pattern, measuring 1 7/8" x 1 1/8":

MANUFACTURED BY
JOSEPH J. WALTON
NEW YORK
HITCHCOCK'S PAT.
OCT. 7. 1879

Comments: Though basically a design insufficiently robust for combat, the Hitchcock Patented Box was compact and allowed the insertion or removal of 20 rounds without necessitating opening the sealed ammunition box. The Army used the same concept in its early (black) M1896 Pistol Cartridge Box when used in conjunction with .45 revolver ammunition.

Roswell D. Hitchcock, Jr., of Washington, D.C. filed for a patent on "certain new and useful Improvements in Cartridge-Boxes" on Sept. 17, 1879. On Oct. 7, 1879, Hitchcock was granted Patent No. 220,377. Hitchcock's patent application, quoted as follows, clearly describes the box patented (shown in Photo 1) as well as the cartridge box actually produced:

"Be it known that I, ROSWELL D. HITCHCOCK, Jr., of Washington, in the county of Washington and District of Columbia, have invented certain new and useful Improvements in Cartridge-Boxes; and I do hereby declare that the following is a full and exact description there-of, reference being had to the accompanying drawings, making a part of this application."

"My invention relates to certain improvements in cartridge-boxes. It has for its object to provide the box with a lid which is readily opened, and at the same time adapted to be folded underneath the bottom of the box to get it out of the way, and when so folded to serve as a brace to prevent the box from tilting outwardly; and, with these ends in view, my invention consists of a cartridge-box having the lid or cover so attached thereto at one end that it can be, when opened, swung outwardly, downwardly, and underneath the bottom of the box, and secured in that position, for the purposes hereinafter more fully and in detail explained."

"My invention further consists in the peculiar construction, hereinafter described, for carrying out the broad idea embodies in my invention."

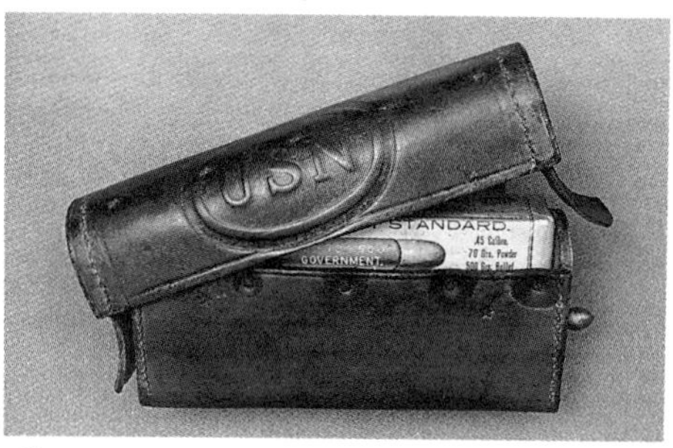

Photo 2.

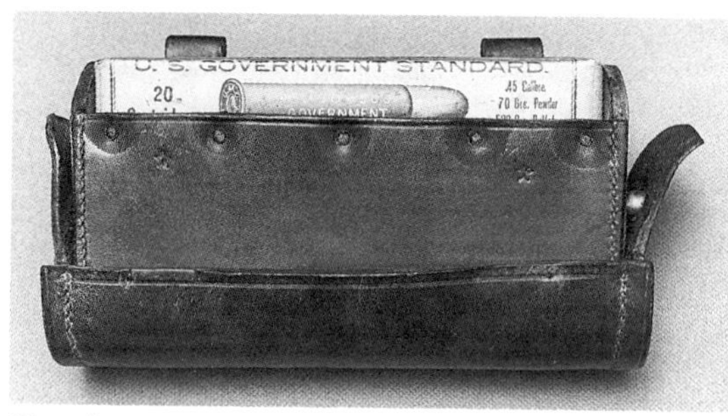

Photo 3.

"Prior to my invention all cartridge-boxes with which I am familiar have been so organized or construct;ed that the lid, when open, either requires to be held in such open position by the wearer, or if designed to keep itself open by spring connections or gravity, the said lid projects from the front or end portions of the box in an unsightly manner and greatly to the disadvantage of the wearer. Another great objection to all cartridge boxes at present made rests in the fact that the weight of the contained ammunition tends to tilt the box outwardly from the person of the wearer."

"My improved box is designed to overcome all these disadvantages; and I have shown in the drawings the best means known to me of carrying out my invention."

Photo 4.

Photo 5.

Carbine Boots

Overview

Because the story of the development of the means for carrying the carbine on the military saddle has not been explicitly told for the period of the Dragoons through the adoption of the true carbine scabbard, the author has chosen to extend his coverage of this subject back in time to the 1840's.

In the mid-1840's (the Mexican War), the longarm of the U.S. Dragoons was either the Model 1847 Musketoon or a Hall Carbine. When mounted, the rider carried this carbine or musketoon either snapped onto the sling swivel or looped at the small of the stock (or trigger-guard) with a leather strap attached to a pommel ring on the near side. The former type of carry (on the snap swivel) was used when action was anticipated and the latter only when an uneventful march would allow the weight of the carbine to be moved from the rider's shoulder to a support strap on the saddle itself. In either mode of support, regulation required the use of a "Carbine Strap and Boot." The "Boot" was, in actuality, a leather muzzle cup, about four inches long, that attached to a 28-inch leather pommel strap about 3/4" wide. This strap buckled to the forward pommel ring on the near side, crossing over the pommel to the off side, where the leather cup was thrust forward by the muzzle of the carbine or musketoon. This arrangement is well illustrated on p. 163, Vol. II of Steffen's The Horse Soldier, and p. 56, Steffen's United States Military Saddles.

The strap and boot method of carry kept the carbine lying high across the rider's right thigh, where it would constantly rub or bounce on the march. When the carbine was snapped onto the carbine sling's snap swivel, most of the carbine's weight was transmitted to the rider's left shoulder, via the wide leather shoulder sling. This problem was not to be relieved until the Army's adoption of a supportive carbine "boot" of an entirely different design.

Regardless of the type of saddle in use, until the 1859 adoption of the McClellan Saddle, carbines were carried in the early style carbine strap and boot, i.e., a simple leather muzzle cup and strap arrangement, with the weight of the carbine usually borne by the rider's left shoulder, via the carbine shoulder sling and snap swivel.

In 1859, Army Ordnance and the cavalry adopted the saddle designed by George Brinton McClellan, Captain of the First Cavalry. With this new saddle came a number of new equipments and concepts, one of which was the method for carrying the carbine while mounted. Dropping the old muzzle "strap and boot" from the Dragoon era, the new concept was considerably simpler, but no less uncomfortable for the rider. The McClellan system included a simple cylindrical, leather socket or "thimble", buckled to the girth strap ring on the off side of the saddle. It became regulation that the cavalryman would leave his carbine snapped into the swivel of his carbine sling at all times and that, while mounted, the carbine would hang behind his right leg with its muzzle thrust through the thimble on the quarter strap ring.

This method of carry was simple enough, and did relieve the rider from having his carbine thrust forward and over his right thigh (as in the old strap and boot system), but the new system did nothing to secure the butt and receiver of the carbine–which increasingly bounced and swung as the gait increased. Further, the full weight of the carbine was still on the rider's left shoulder, via the shoulder strap. Neither of these problems were to be relieved for another 25 years . . . with one exception.

The Sumner Boot

In March of 1878, Captain Samuel Stroow Sumner, 8th Cavalry, stationed at Presidio (San Francisco), California, was given Ordnance approval for the manufacture of a sufficient number of carbine "sockets" to supply his company on a trial basis. These "sockets" were made at Benecia Arsenal and the cost was 90¢ each. On 29 March 1878, one specimen was submitted to the Ordnance Office "for remark."

Sumner's idea was the genesis for the short and long carbine boots adopted in the 1880's, but its reception in Washington was typically critical and somewhat political. First, the socket (actually, a "boot" in modern collector terms) was not popular with General Sheridan, who deplored the likelihood of "burdening the poor horse with the dead weight of the carbine." Assumably, the general felt that horses were more valuable than soldiers, and that carbines were somehow lighter when attached to the rider's carbine sling. Second, the Ordnance Office stated that they had no complaints with the current sling and socket system, c. 1859. One cannot help but wonder just how closely the Ordnance officials on the Potomac were monitoring actual field needs on the Frontier–a point reflected upon by the Commander of Benecia Arsenal, later in this discussion. Third, Colonel Hagner himself seemed to choose to misunderstand that Sumner was suggesting that the new "socket" was being offered as a substitute for the carbine sling and swivel. Such was not the case, but then all of those career officers could not help but remember that General Sheridan didn't like the idea.

Sumner's voice was not raised alone, however, in support of his idea. The Commanding Officer of Benecia Arsenal, in one correspondence to the Ordnance Office in behalf of the new idea, commented, "The Cavalry officers in this division report that the sockets do not answer the purpose for which they were made." He went on to note that, "the ideas, in reference to equipment, of officers serving in the field and who have to make long marches, may differ from those who never mount a horse without being considered faulty." Somewhat later, this same Ordnance officer commanding Benecia, vainly called the Ordnance Office's attention to "the fact that all packers consider it necessary for the comfort of their animals to fasten on their packs as rigidly as possible." (Offering field experience against the currently jolting, dangling carry of the old carbine sling and thimble system.) To no avail. Washington didn't like new ideas, particularly from the field.

The Hartman Socket

In late 1878, the Secretary of War convened a special board of officers to review Army equipments of specific types, and in general. One of many recommendations made in 1879 by this board was that a carbine socket be retained, but that it be a different type of socket, one designed by Sergeant Henry Hartman, First Cavalry. Sgt. Hartman had originally submitted his improved, "split-circle" socket idea to the Ordnance Office in January, 1876. The idea, sketches and a sample made at Benecia, were submitted with very favorable recommendations by both Hartman's company and regimental commanders. The archival records are incomplete, but do show Watervliet Arsenal's typical criticism of the new socket idea on the basis of increased cost. The idea did not die, however, and the Hartman Socket, to quote the 1879 Board, was one with "decided advantages over the present patterns, in the facility with which the carbine can be liberated by simply pressing the carbine outward, a great safeguard were a rider to be thrown with his carbine slung." Interestingly, the socket concept was being continued, but the Board also recommended a carbine loop be adopted for the pommel of the recommended Whitman saddle, "believing that to be the most convenient place to carry the carbine on scouts and long marches." The Board went on to note that, while many Frontier civilians carried their longarms in this manner, "in many cavalry regiments company officers permit their men to carry their carbines in a similar way." The Ordnance Office, upon submission of these and many other recommendations, non-concurred with the pommel loop idea (by disapproving the Whitman Saddle) but did not disapprove the Hartman Socket concept. General Sherman, General of the Army, however, placed greater value on the recommendations of the Board's "five officers, selected by reason of their special fitness and their large experience with the wants of the troops in actual service on the plains and in the mountains of the interior of the continent," and recommended to the Secretary of War that most of the proposed Board recommendations be approved and some phased in over time–including the Whitman Saddle and its carbine pommel loop concept.

Still, the field cavalry was a long way from relief from the old sling and socket carry of the carbine. Only as the thousands of Civil War surplus McClellans were diminished by use were the new Whitman Saddles to be manufactured and issued. (In fact, this never occurred in quantity. The Hartman Socket, however, was produced in limited quantity and used until it was finally replaced by a true carbine boot, five years later.)

The 1885 Carbine Boot (The "Short Boot")

In 1884, another board of cavalry officers was convened to review cavalry equipments, current and proposed. One of their recommendations was to drop the Hartman Socket and replace it with a carbine boot very similar to that proposed six years earlier by Captain Sumner. Sumner, now a cavalry major serving in Texas with the 8th Cavalry, must have smiled with some irony at this decision. The Board's recommendation for a boot was accepted and what collectors now call the "Short Boot" was adopted.

The new boot was suspended from the saddlebag stud by a single 17 1/2" strap, punched with two adjustment holes (to adjust for the rider's preference). It was attached to the off-side rear girth strap and/or the girth strap ring by the upper and lower boot straps.

Initially, the two straps for attachment of the boot to the rear girth strap and the girth strap ring were arranged with the upper strap about mid-point on the boot body and the lower strap secured with one rivet around the lower end of the boot. Since riders and horses differ, the actual attachment of the new carbine boot did not suit all riders. This led to the occasional, unofficial movement of the lower boot strap up toward the upper strap–usually by 1 1/2 to 2 inches. However, the original spacing of the two straps on the boot body never changed during arsenal manufacture of both the Short Boot and the later Long Boot.

As shown in the original Ordnance drawing of 1886, the original point of attachment of the Short Boot's saddlebag suspension strap was back near the stitched edge of the boot. This kept the carbine in a more or less vertical position–and placed the butt of the carbine too far forward for the rider's easy access to his revolver holster. And, for some riders, it may have allowed a constant bumping of his right elbow and arm. Perhaps for both of these reasons, many of the suspension straps were moved four inches forward on the boot. This allowed the boot to rotate slightly backward and moved the carbine butt further to the rear. This new position for the suspension strap was the one adopted in the "Long Boot" (about 1887).

A final note on the original manufacture of the Short Boot will come as a surprise to some collectors. Simply put, it appears that a large number of Short Boots were incorrectly assembled at Watervliet Arsenal and required modification before they were in regulation configuration and could be issued and used. Examination of almost all Short Boots today (and those modified to Long Boot configuration) shows this corrected manufacturing error.

It is clear from the artifactual evidence that the early Short Boots were <u>originally</u> assembled with the saddle bag stud suspension strap on the outside rear of the throat. As odd as this seems, careful inspection of most Short Boots will show a small, arsenal-plugged rivet hole just opposite the commonly-seen placement of the suspension strap rivet. (Photos will be shown later in this section.) Inspection of the inside of the boot on many specimens will also show that the hole actually had a burred rivet in it, indicating that the suspension strap was assembled on many, if not all, of these boots. The most plausible explanation of this feature is that a large number of Short Boot blanks were punched in error and actually assembled before the error was caught by inspectors. Such boots actually fully assembled (with the suspension strap reversed) would not have been satisfactory for field use–as some reflection on the problems of mounting and using such a boot will quickly indicate.

[NOTE ON THE 1884 ORDNANCE BOARD AND THE DRAWINGS IN ORDNANCE MEMO. NO. 29: Most cavalry collectors own a copy of Ordnance Memorandum No. 29 and know that the carbine boot shown on Plate VIII is not the one discussed above, but is the later, 1887-era "Long Boot". The reason for this is that the 1884 Board's recommendation of the Short Boot was adopted in 1885. Though some production of this boot probably took place in 1884, it would be 1886 before Rock Island Arsenal had produced a set of drawings to accompany the Board's recommendations and the eventual Ordnance memorandum. By the time the Ordnance Memo. was actually published in 1891, the Long Boot–to be discussed below–had been adopted and drawings and descriptions in the draft Ordnance Memo. had been changed to reflect the latest pattern of carbine boot (the Long Boot, adopted about 1887).]

The 1887 Carbine Boot (The "Long Boot")

Since the 1885 Carbine Boot was the first to see extensive field service, it was inevitable that field service gave rise to requests for changes in the design–as well as field modifications already discussed. The first modification suggested from the field was that the mouth of the carbine boot be flared to more easily allow the rider to more easily thrust the carbine muzzle into it. (More difficult while on the move than the reader might imagine.) A second improvement suggested for the basic 1885 Carbine Boot was to strengthen and protect the forward portion of the boot's throat from the constant abrasion of the carbine's rear sight. Flaring the throat, as suggested earlier, would remedy part of this problem, but it was even more fully resolved with the riveted addition of a curved brass plate, which both protected the throat, but helped hold it in a flared position. The third improvement of the carbine boot was occasioned by the Army's adoption of the Model 1884 Carbine but, more specifically, the somewhat fragile Buffington-pattern rear sight.

When the Board of 1884 first convened, the carbine in heaviest use was the Springfield with the Model 1879 rear sight. This robust arm had a rear sight well-designed for hard service use, and incorporated a set of protective "ears", much like the rear sight of the M1917 Enfield Rifle of WWI fame. The ears on the M1879 rear sight easily protected the sight as it was removed from and returned to the 1885 Carbine Boot. The boot had a swell wet-impressed on the front of its body to accommodate the M1879 rear sight. As the 1884-approved Buffington sight began to appear, it was apparent that the original boot design of 1884 would not accommodate it well. The front edge of the 1879 sight was 4 3/4 inches from the receiver ring. The front edge of the 1884 sight was 5 1/2 inches from the receiver ring and that allowed it to protrude 1 1/4 inches beyond the end of the 1885 Boot. Ever frugal, the Ordnance solution was to modify current arsenal inventory of 1885 boots with a flared and brass-reinforced throat and to carefully splice (onto the lower end of the boot) a 2 1/2 inch "lower boot section". To satisfy the new rear sight requirement, while the boots were split for the modification process, a new sight swell was wet-pressed onto the forward edge of the boots. The new swell extends 1 3/4 inches down into the spliced section.

The last improvement for the new carbine boot was to accept the field's obvious desire for a design that would tip the throat of the boot back somewhat–and the suspension strap was moved forward four inches to accommodate this requirement.

The final carbine boot, the 1887 Long Boot, just as its predecessor, was manufactured solely at Rock Island Arsenal. It seems to have seen extensive production and service until it was finally replaced in the early 1890's by the Army's first carbine scabbard. Interestingly, this new scabbard retained the brass throat plate of the 1887 Boot, but the carbine's placement was moved to the front near side with the butt forward and the trigger-guard up. Still, the carbine sling and swivel were retained for several more years.

Note on terminology of the Indian War period: Using the Sumner Carbine Boot and the Hartman Socket as good examples of the ambivalence of terminology for the new ideas, the following give some idea of the difficulty in following these articles through time.

Sumner Carbine Boot: Benecia Arsenal originally called this a "Carbine Bucket" on its submission to Washington. The Ordnance Office referred to Sumner's Boot as "this form of socket", in contrast to the "old carbine socket" (M1859). In the same 2nd Endorsement letter from Watervliet Arsenal, the Sumner Boot was called a "boot" and the M1859 Socket, a "socket." And, finally, by the 6th Endorsement from Watervliet Arsenal, the Sumner Boot was called "a boot".

Hartman Socket: Sgt. Hartman himself called his socket a "carbine boot" in his original 1876 submission up through channels. Benecia called it a "socket" in their 2nd Endorsement. Watervliet Arsenal, in their 4th Endorsement, called the concept a "boot."

What this suggests to the student of military history is that one must be careful to have some perspective and flexibility when researching these early cavalry developments. The same may be said of the occasional interchangability of the terms, "pouch," "box," "pattern," "type" and "model".

1859 Carbine Thimble

Adopted in 1859 as part of the 1859 McClellan Saddle equipment, the carbine thimble (or "socket", as it soon came to be called) was a departure from the previous "Carbine Strap and Boot" used for years by the Dragoons. Simpler and less costly to produce, the new thimble was a wasp-waisted leather cylinder 2 1/2 to 2 3/4 inches high and approximately 2 1/2 inches across. A single strap buckled the thimble to the off side girth strap safe, making it a permanent part of the saddle.

The carbine barrel was inserted through the socket for a distance of from 4 to 8 inches, and was held there, laterally, until withdrawn. The weight of the carbine, however, was always supported by the carbine sling over the cavalryman's shoulder and attached to the carbine by its ring.

The thimble was made by many makers during the Civil War, and the names and, occasionally, the city and state of the maker were stamped on the tip of the strap. The single strap buckle is a flat, japanned iron buckle with beveled edge. There is considerable variation in the sockets made during the Civil War, and when the War ended, the great surplus of these sockets continued to serve the cavalry until they were phased out in the late 1870s.

Photo 1.

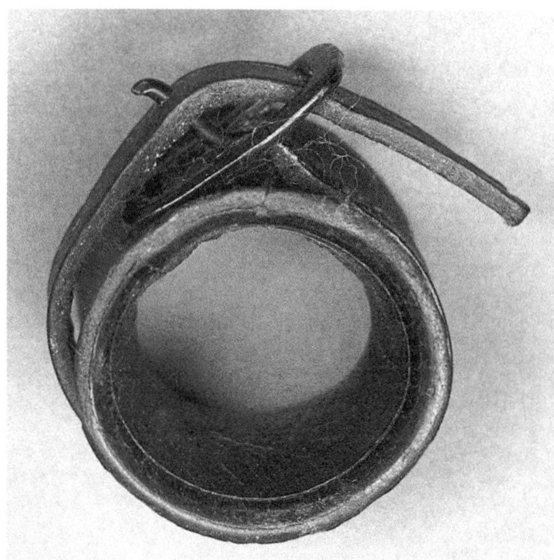

Photo 2.

The Sumner Carbine Boot

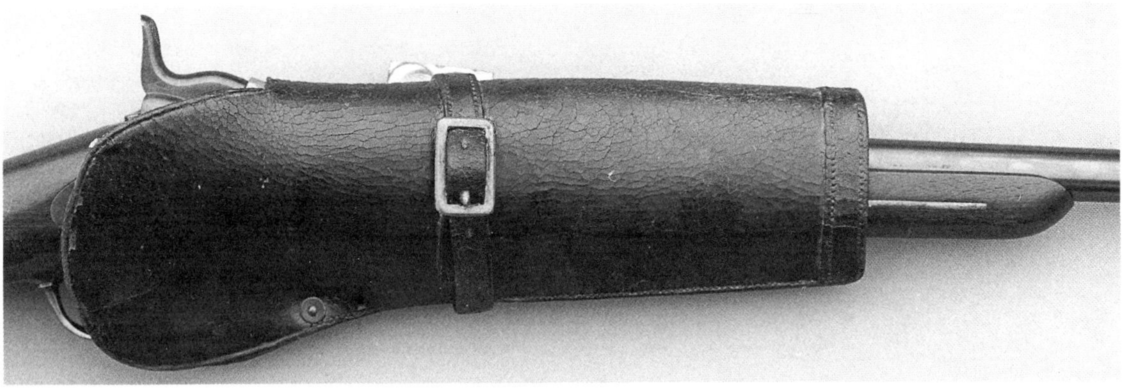

Photo 1.

Photo 2.

The carbine boot designed by Captain Samuel Stroow Sumner of the 8th U.S. Cavalry in 1877-78 was the genesis for the carbine boots adopted for the cavalry in 1885 and about 1887. Typically, the Ordnance Office never recognized Sumner's contribution, but adopted the concept as their own, albeit grudgingly, under pressure from the field commanders.

At least one troop of 8th Cavalry was outfitted with the Sumner Boot on a trial basis in 1878 and the results must have been so noteworthy as to cause field cavalry troops and commanders to remember its use and continue to press for a true "boot" over the five-year period following its Far Western trial.

The specimen shown is the only known specimen of the Sumner Boot, and was found still carrying the 1878 Ordnance Office property tag marked, **"Carbine Bucket from Capt. Sumner's Company Presidio–1602 Ordnance Office Rec'd Mar. 29 1878."** It is, truly, the rarest of all carbine boots of the Indian War period.

The boot is manufactured of black bridle leather that is somewhat lighter than the later, regulation boots. It has three brass fittings: a box buckle for the body strap, a 1 5/8 inch-diameter ring joining the suspension strap to the chape on the boot body and a wire buckle for length ad-

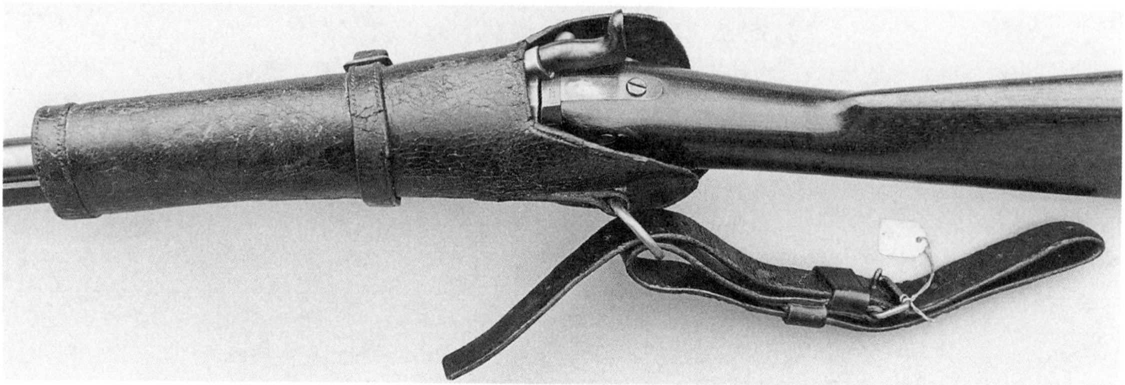

Photo 3.

justment of the suspension strap. The brass ring allows the same movement of the upper boot body as the single rivet adopted for later boots.) The lower end of the boot is reinforced with an 11/16 inch-wide strap, sewn to the body. The upper end of the stitched seam of the boot body is reinforced with a single rivet, as shown in Photo 1. Photo 2 shows the full length of the boot on its near or inside. Photo 3 is a view of the top of the boot, showing the square cut at the top of the opening.

Full length of the Sumner Boot, with strap, as shown in Photo 2, is 23 1/4 inches. Length of the boot body is 13 inches.

The 1 1/16 inch wide suspension strap is quite long and could, actually, be attached to either of the cantle rings, but is much more appropriately attached to the ring on the off side. The strap is perforated at 1 1/2 inch intervals for much of its length, giving greater adjustment variation.

The chape securing the suspension strap to the body via a brass ring is both stitched to the body and secured with one rivet. Note: the interior of the rivet has a domed head to avoid scratching the carbine.

Hartman Carbine Socket

The short, somewhat wasp-waisted carbine socket or thimble adopted by the cavalry as part of the M1859 McClellan Saddle equipments, had one major flaw–on occasion, it got its rider killed in a most unpleasant manner.

Regulation, from at least 1859 until the late 1890's, required that the cavalry soldier keep his carbine attached to the carbine sling with the large, swiveling snap on the sling. This sling supported the carbine totally, while the muzzle was restricted in movement by the circular carbine socket. Normally, proper use saw the carbine barrel and front sight protruding through the socket at least 8-9 inches. Under normal circumstances, this was a low-weight, cost-efficient approach to carrying the carbine, but it quickly became lethal for the rider if he was thrown and was unable to disengage either himself from his carbine sling (over his left shoulder) or his carbine barrel from the socket. The hapless rider, thrown while the carbine was in place and slung, stood a good chance of being dragged to death by a spooked horse, or trampled by his or other mounts.

The cavalry and the Ordnance Dept. seem to have accepted these occasional accidents as acceptable mortality, though this type of accident was well known in the service.

In January 1876, Sergeant Henry Hartman submitted an idea for an improved carbine socket up through Army channels. A model socket, manufactured at Benecia Arsenal, was also submitted. Hartman's idea was a simple one, but was one that could save lives. It consisted of a leather-covered spring steel socket, open slightly on the front side. Under normal circumstances, the socket would retain the carbine barrel, but if the rider wished to remove the carbine, all he had to do was grasp the butt or small of the carbine and tilt it forward against the opening of the socket. The socket would spread and release the carbine instantly. Even more importantly, if the rider was thrown, the secured carbine would spring the socket and fall free with the rider.

It is unknown what the Ordnance Office felt about Hartman's life-saving idea, but Watervliet Arsenal noted it would be 21.65 cents more expensive than the current carbine socket and seemed to miss, entirely, the advantage to a thrown rider.

Hartman's idea did not die, however, and in 1879 the Cavalry Board recommended the Hartman Socket as a replacement to the M1859 Socket and the Hartman Socket was approved as cavalry standard. The socket served as a regulation item until replaced by the 1885 Carbine Boot (Short Boot).

The Hartman Socket is simply a large, leather-covered spring steel loop that is open 3/16 inch at the front edge. Upper and lower edges are markedly flared. As can be seen from the photo, the attaching strap is riveted to the socket body in two places, looping back on itself to buckle on the standard box buckle. The specimen shown is unmarked and measures as follows:

OAH–	2 3/4"
OAW–	3"

Hartman Sockets today are rare–despite the fact that they were regulation for five years. This suggests that they did not see full cavalry issuance before being replaced by the 1885 Carbine Boot.

1885 Carbine Boot-First Pattern

The drawing (Photo 1) here shows the 1885 Carbine Boot, as adopted and manufactured. As discussed in the Introduction to this section, the delay of two to three years before all the drawings for Ord. Memo. No. 29 were completed gave time to the Army to develop a better pattern of carbine boot and the drawing of the later 1887 Carbine Boot was substituted before the Memo. was actually published in 1891.

The design of this boot is a strengthened and somewhat simplified version of the Sumner Boot pattern. The leather used is a much heavier bridle leather. An additional body strap has been adopted to better secure the boot to both the girth strap and girth ring. The 17 1/4 inch suspension strap has been secured to the upper boot body by a single brass rivet–which allows full rotation around its pivot. The boot body measures 11 1/4 inches in length. Both buckles on the body straps are japanned iron and the straps themselves are riveted to the boot body. (The upper strap in two places and the lower in only one.)

The First Pattern suspension strap was riveted on the near side, close to the stitched seam. The regulation position of the lower body strap is on the lower end of the boot, though this was often moved in field service.

Photo 2 shows the near side of an actual regulation 1885 Short Boot. The two lower holes in the strap are the regulation holes and are spaced 2 inches

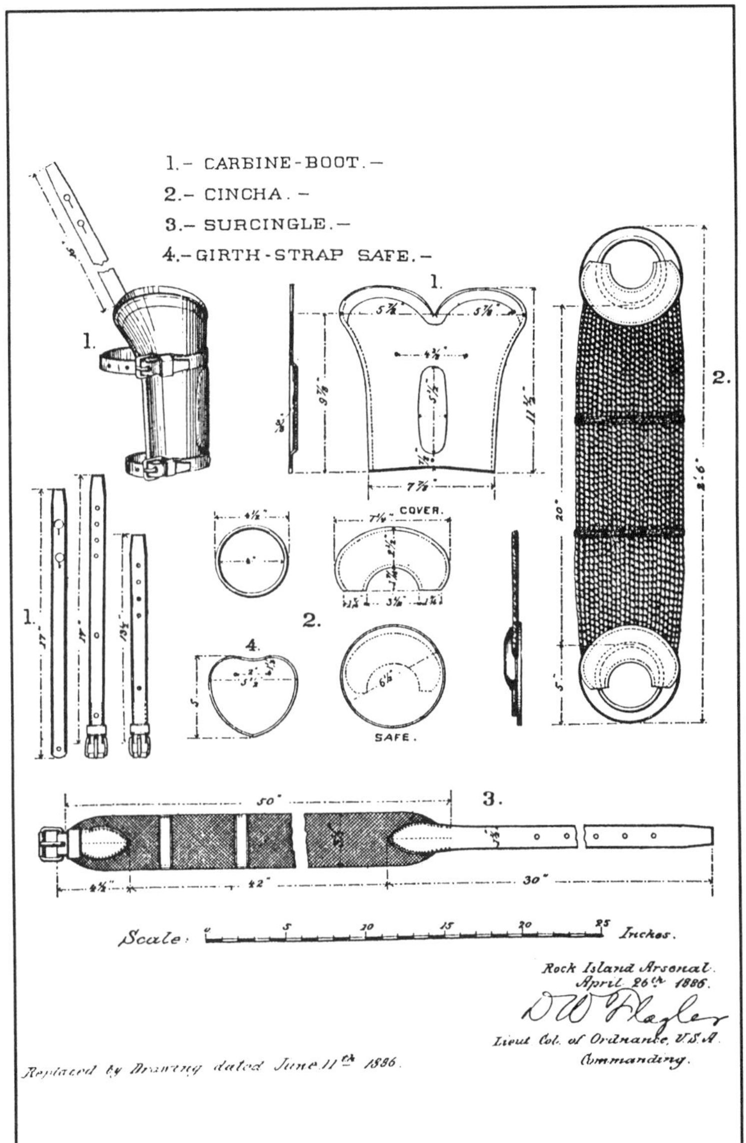

Photo 1.

apart. The suspension strap hole nearest the boot body is a field modification, and would have raised the boot and carbine another 3 1/4 inches.

The pointer in Photo 3 shows the arsenal-plugged hole that most 1885 Boots examined have (though in some, the plug has fallen out over the years.) All

1885 Boots shown in this book have this plugged hole. The only reasonable explanation for this consistent feature is that a mistake was made at Rock Island Arsenal, possibly in the cutting templates used, and the saddlebag stud suspension straps were actually assembled on many of the boot bodies before the mistake was identified. The carbine boot could not be properly used with the suspension strap in this position, and all assembled boots were probably plugged and the suspension straps moved before issue.

It is interesting to note that, with the adoption of the 1885 Carbine Boot, the cavalry dropped its life-saving, quick-release Hartman Socket and returned to a much more dangerous method for carrying the carbine. That is, since regulation required the carbine to be snapped onto the carbine sling even while in the carbine boot, a rider thrown from his horse had a very good chance of being dragged to serious injury or death. This problem was continued in the cavalry until the later 1890s, when the carbine sling was at last phased out of service.

This particular boot is marked **"F 6 11"** for F Troop, 11th Cavalry, #6 (an issue number).

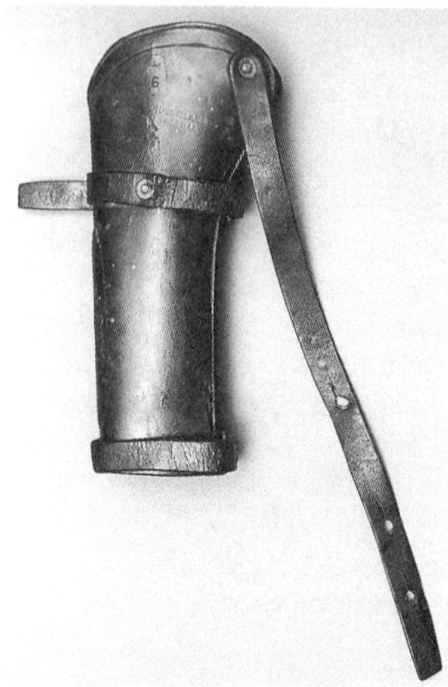

Photo 2.

Photo 3.

1885 Carbine Boot, First Pattern– Field Modification

With an 1879-era "Trapdoor" Carbine in it, this boot shows a typical early field modification, whereby the lower body strap has been moved up the body and re-riveted. The suspension strap, however, is in its original position back near the stitched seam.

Note that, invariably, when the lower strap was moved up it was re-riveted with two rivets (one on each side of the body). The strap's original method of attachment, in its lower position, was with only one rivet in the lower front of the boot.

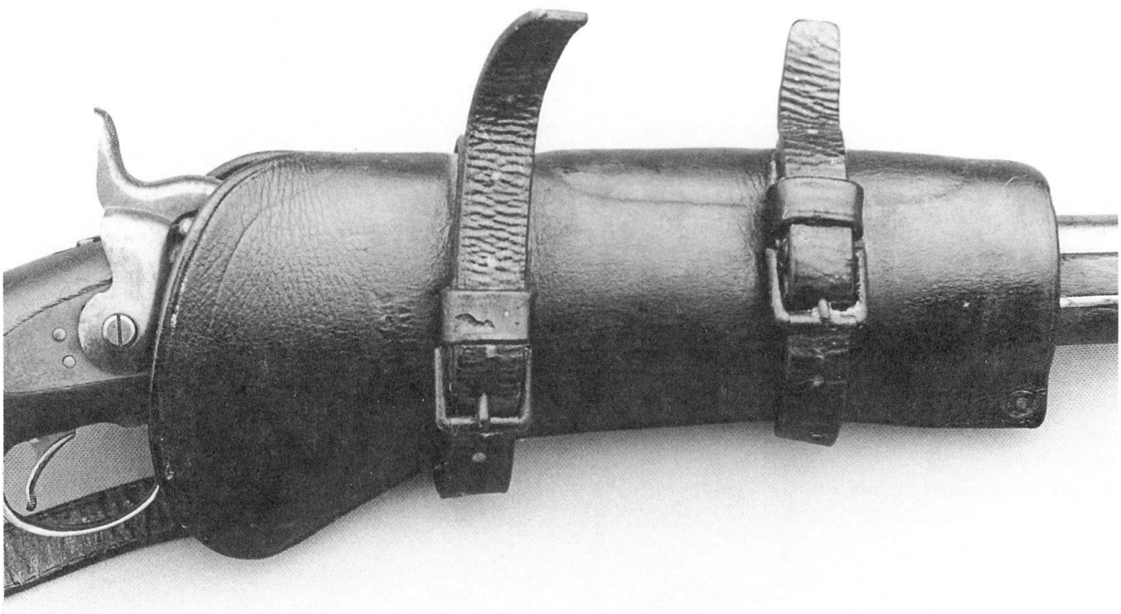

1885 Carbine Boot–Second Pattern

From the artifactual evidence in collections today, it appears that the saddlebag stud suspension strap placement was unacceptable to most users. In its original (First Pattern 1885 Boot) position, the boot was held in a more or less vertical position. This placed the butt of the carbine in a range of interference with the rider's right elbow, arm and hand. To move the butt more out of the way, early boots were modified by moving the suspension strap rivet several inches forward, imparting a rearward tip to the mouth of the boot and moving the carbine butt six inches or more to the rear. This troop preference was noted by the Ordnance Office and subsequent production of the 1885 Boot placed the suspension strap four inches forward of the position adopted in the First Pattern 1885 Boot. These Second Pattern Boots will not have a rivet hole in the earlier position.

The accompanying photo shows a Second Pattern Boot with the suspension strap moved well forward on the upper boot body.

This particular boot is marked **"G 12 10"** for G Troop, 10th Cavalry, #12 (an issue number). The 10th Cavalry was one of two regiments of "Buffalo Soldiers."

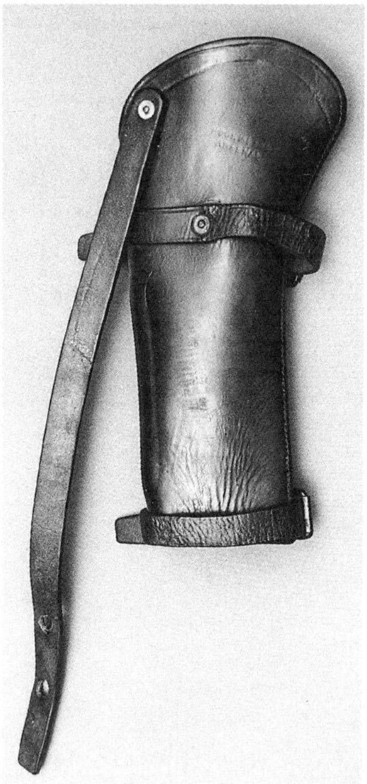

1885 Carbine Boot, Second Pattern–Field Modification

This Second Pattern Boot shows a fairly common field modification, wherein the lower body strap was moved 1 3/4 inches up the body. This was probably done to accommodate some trooper's particular idea of comfort or convenience. Such a modification would require that the lower strap no longer attach to the girth strap ring but would, like the upper body strap, attach around the right rear quarter strap.

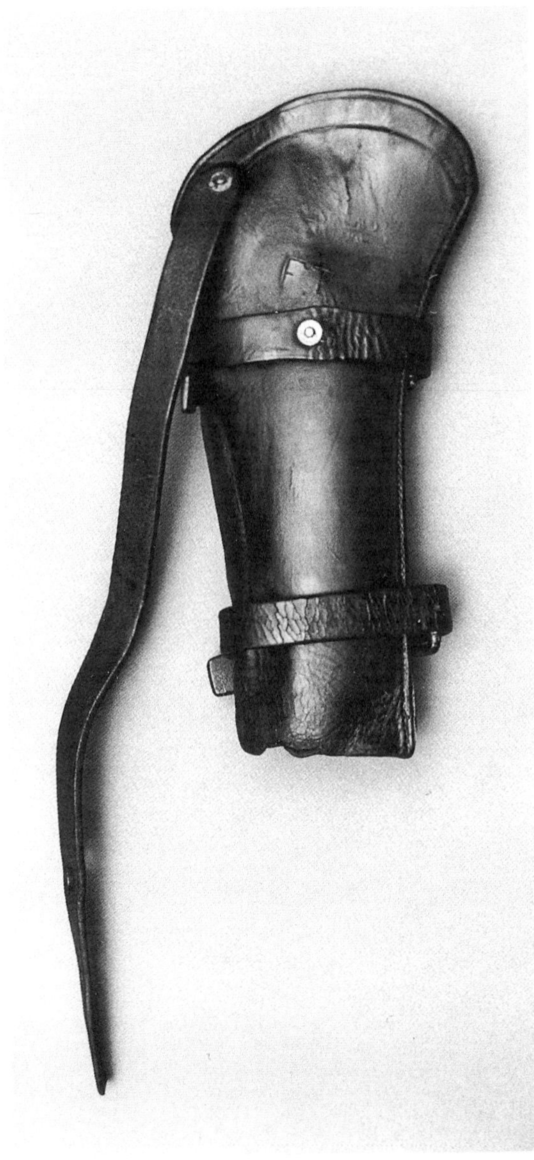

1885 Carbine Boot, Second Pattern—Field Modification

At some point in its service life, the suspension strap was moved to a mid-point on the throat, and it and the lower body strap were re-riveted. Later yet in the boot's service life, the suspension strap was replaced in its original position and the lower body strap was re-riveted in its original position. To accomplish this, rather than grinding off the rivet heads and punching out the rivets, a circular cutter was used to cut out both rivet and burr (much like a wad cutter for a shotgun wad).

The two arrows in the photo indicate the two holes left by the cutter when the rivets and burrs were removed.

The "Extended" Boot

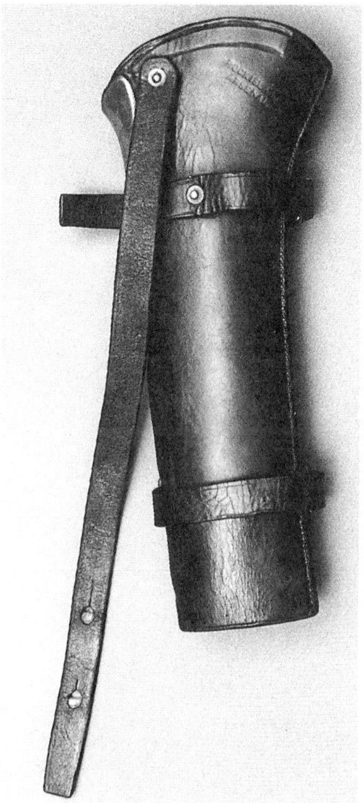

Photo 1.

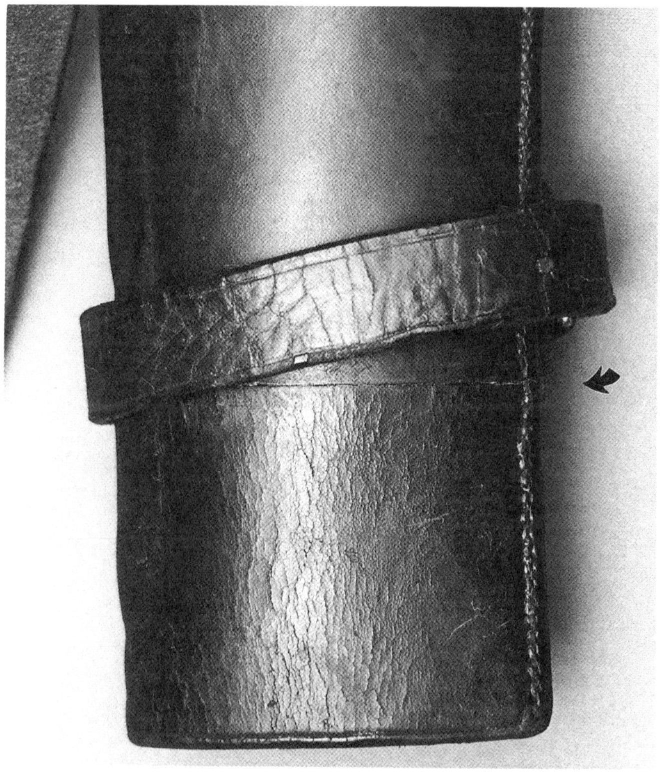

Photo 2.

As discussed in the Overview of this section, the years following 1885 gave the cavalry and the Ordnance Office opportunity to consider improvements to the carbine boot in the service. Improvements suggested were for a boot more flared at the mouth to ease entry of the carbine muzzle, a strengthened mouth or opening and protection of the fragile and exposed rear sight on the M1884 Springfield Carbine.

It is not known just when these improvements were approved by the Ordnance Office, but it could have been as early as 1887. However, many of the 1885 Carbine Boots remained in stock and these were probably earmarked for conversion to the new pattern before the new pattern boots were manufactured from scratch.

The conversion of the older, Short Boots to the new design required several changes: 1. Extension of the boot body by 2 5/8 inches. This was done by opening up the boot and sewing in an extension on the lower end. 2. Flaring the mouth of the boot to ease the entry of the carbine. This was done by cutting the mouth of the boot and inserting a triangle of bridle leather to hold the mouth in a permanent flare. 3. Reinforcement of the mouth of the boot with a flared brass plate, riveted in four places. 4. Protection of the Buffington Sight (which protruded 13/16 inches further down the barrel than the old 1879 Model Sight) was accomplished by wet-pressing another, lower sight swell on the boot body. This left a two-stage sight swell on the body, as shown in Photo 4.

These extended boots conform exactly to what is now

called the 1887 Boot specifications and can only be identified by lifting the lower body strap to view the extension seam (Photo 2). Photo 3 shows interesting details from the conversion process. The top arrow shows the unplugged hole from the attachment of the suspension strap when the boot was configured as an 1884 First Pattern. The lower arrow points to a misplaced rivet hole, made for the brass throat plate during the conversion process. (There is another such hole on the other side.)

Photo 4 shows the off side of the converted boot. The long body strap tips shown here are often found trimmed back by cavalry troops. In practice, the overly-long straps flapped and, probably, caught the rider's leg on mounting and inserting his foot into the right stirrup.

A final comment should be made on this intriguing and scarce carbine boot. It is currently impossible to say whether this boot preceded by some time the production of the 1887 Carbine Boot or whether it was done at the beginning of the 1887 Boot production. Either is a possibility. However, the fragility of the 1884 Carbine rear sight must have spurred the Ordnance Office to provide a more protective carbine boot than the 1885 Short Boot. And, that office went so far as to design and issue a special barrel band with an integral sight protector to partially solve the sight damage problem. In this context, the author considers it reasonable to theorize that the "Extended" Boot may have been produced as early as 1886.

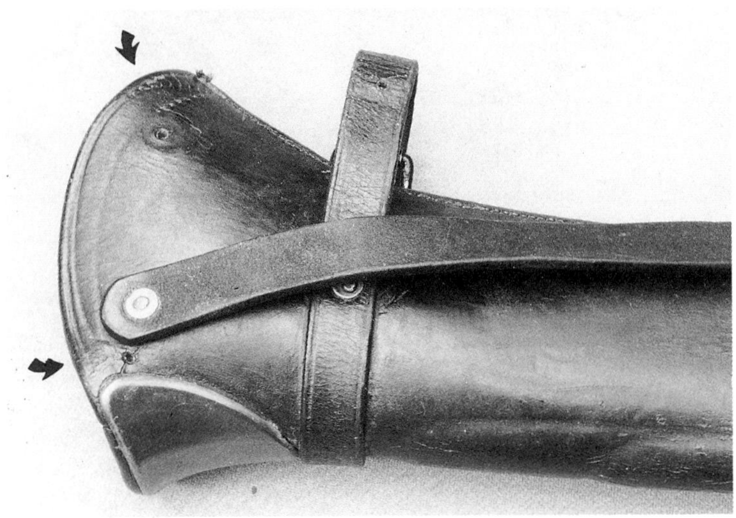

Photo 3.

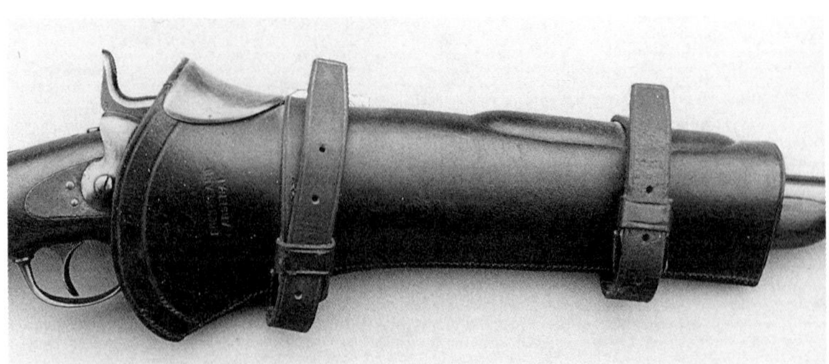

Photo 4.

1887 Carbine Boot

The true 1887 Boot incorporated all of the improvements of the period and served the Regular Army cavalry until the .45/70 Springfield Carbine was replaced in 1896 by the .30 calibre Krag Jorgensen Carbine. However, this boot continued in service in great quantity in the many state militia cavalries into the late 1890s.

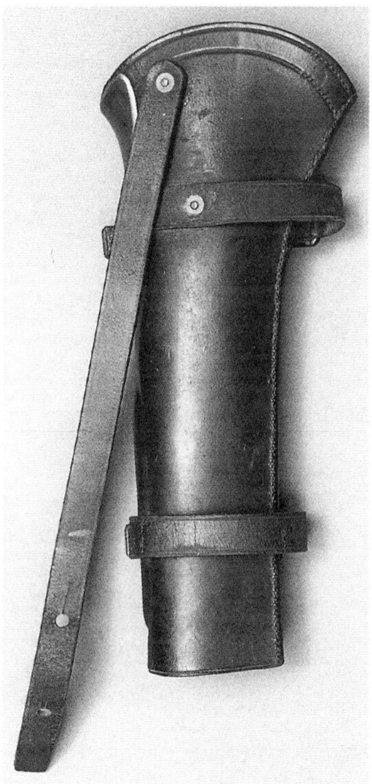

Photo 1.

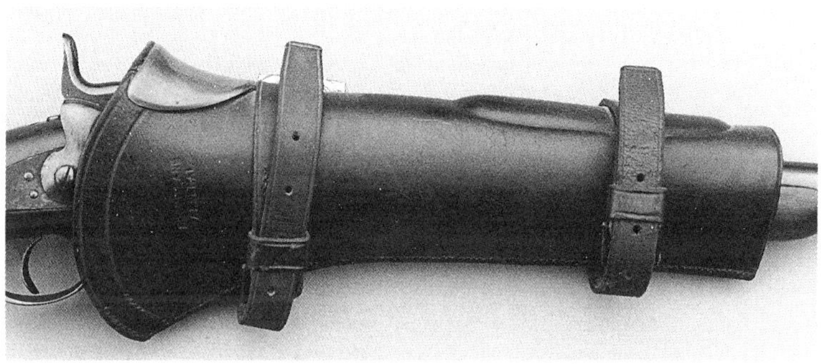

Photo 2.

1887 Carbine Boot-Field Modified

An 11th Cavalry-marked boot showing yet another type of modification for a real or imagined advantage to the rider.

In this case, the lower body strap has been reversed to buckle on the back–presumably around the quarter strap ring. To mount or remove this boot, the trooper would have to reach around the boot and unbuckle the strap "blind", so to speak. But, since this was not a daily event, the inconvenience may have, somehow, been worth it.

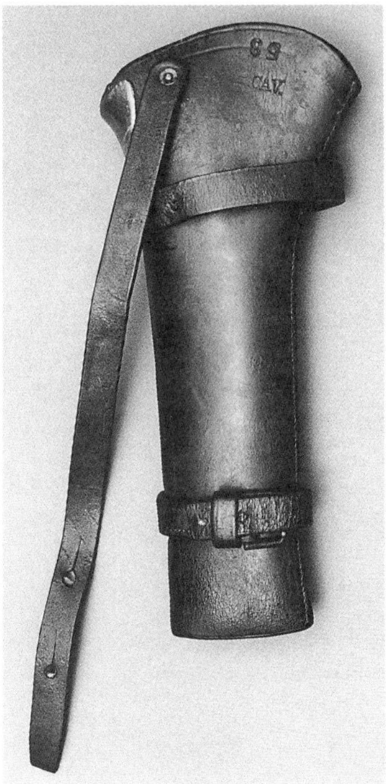

References

1. **Ordnance Memoranda No. 18, Horse-Equipments, Cavalry Equipments, Accoutrements, Etc.**, Government Printing Office, Washington, D.C., 1874.
2. **Ordnance Memoranda No. 19, Infantry Equipments and Materials, Etc.**, Government Printing Office, Washington, D.C., 1875.
3. **Ordnance Memoranda No. 29, Horse Equipments and Cavalry Accoutrements**, Government Printing Office, Washington, D.C., 1891.
4. **U.S. Military Holsters And Pistol Cartridge Boxes**, Dr. Scott E. Meadows, Taylor Publishing Co., TX, 1987.
5. **American Military Belts and Related Equipments**, R. Stephen Dorsey, Pioneer Press, TN, 1984.
6. **Civil War Collector's Encyclopedia**, Francis A. Lord, Castle Books, NY, 1963, 1965.
7. **Bugles, Banners and War Bonnets**, Ernest L. Reedstrom, Caxton Printers Ltd, ID, 1977.
8. **The Horse Soldier 1776-1943**, Volumes I-III, Randy Steffen, University of Oklahoma Press, OK, 1977.
9. **American Military Equipage, 1851-1872**, Frederick P. Todd, Charles Scribner's Sons, NY, 1980.
10. **Bullets or Bacon–A Study of the Dyer Cartridge Pouch, 1870-1879**, J. Edward Green III, Man at Arms, Sept.-Oct. 1981.
11. "Answer: Cavalry Sabres", James S. Hutchins, **Journal of the Company of Military Collectors and Historians**, "Vol. VI, No. 1, March 1954.
12. "The Cavalry Campaign Outfit at the Little Big Horn", James S. Hutchins, **Journal of the Company of Military Collectors and Historians**, Vol. VII, No. 4, Winter 1955.
13. **U.S. Patents from 1857-1887**.
14. "Identifying Civil War Cavalry Cartridge Boxes", Paul D. Johnson, **Journal of the Company of Military Collectors and Historians**, Vol. XL, No. 3, Fall 1988.
15. "Notes on the 1874 McKeever", J. Edward Green III, **Journal of the Company of Military Collectors and Historians**, Vol. 29, No. 1, 1977.
16. "Day's Patent Revolving Cartridge Box", William G. Phillips, **Journal of the Company of Military Collectors and Historians**, Vol. 42, No. 4, Winter 1990.
17. **Gun Tools: Their History and Identification**, James B. Shaffer, R. Stephen Dorsey, L. Rutledge, Collectors' Library, OR. 1992.
18. "Cavalry Accoutrements, 1874: The Rare Pieces", William G. Phillips, **Journal of the Company of Military Collectors and Historians**, Vol. 44, No. 2, Summer 1992.
19. Numerous correspondences from the Army Ordnance Office files of the National Archives, Washington, D.C.